D0464070

THE LITTLE BOOK OF BLACK HOLES

Books in the *SCIENCE ESSENTIALS* series bring cutting-edge science to a general audience. The series provides the foundation for a better understanding of the scientific and technical advances changing our world. In each volume, a prominent scientist—chosen by an advisory board of National Academy of Science members—conveys in clear prose the fundamental knowledge underlying a rapidly evolving field of scientific endeavor.

the

LITTLE BOOK

of

BLACK HOLES

STEVEN S. GUBSER
& FRANS PRETORIUS

PRINCETON UNIVERSITY PRESS   PRINCETON AND OXFORD

Published by Princeton University Press,
41 William Street, Princeton, New Jersey 08540

In the United Kingdom: Princeton University Press,
6 Oxford Street, Woodstock, Oxfordshire OX20 1TR

press.princeton.edu

Cover art and design by Jess Massabrook

Library of Congress Cataloging-in-Publication Data

Names: Gubser, Steven Scott, 1972– author. | Pretorius, Frans, 1973– author.
Title: The little book of black holes / Steven S. Gubser
and Frans Pretorius.
Description: Princeton, New Jersey ; Oxford :
Princeton University Press, [2017]
Identifiers: LCCN 2017024783| ISBN 9780691163727 (hardback ;
alk. paper) | ISBN 0691163723 (hardback ; alk. paper)
Subjects: LCSH: Black holes (Astronomy)
Classification: LCC QB843.B55 G83 2017 | DDC 523.8/875—dc23
LC record available at https://lccn.loc.gov/2017024783

British Library Cataloging-in-Publication Data is available

This book has been composed in Bembo Std

Printed on acid-free paper. ∞

Printed in the United States of America

1 3 5 7 9 10 8 6 4 2

CONTENTS

PREFACE

It was September 14, 2015, almost exactly 100 years after Albert Einstein formulated the General Theory of Relativity. Two massive detectors, one in Louisiana and one in Washington, were undergoing final preparations for a science run aimed at detecting gravitational waves. Suddenly and unexpectedly, the detectors' instrumentation recorded a peculiar chirp. If we made this chirp audible, it would sound like a faint, low-pitched thump.

Five months later, after careful scrutiny of the data recorded by these detectors, the Laser Interferometer Gravitational Wave Observatory (LIGO) publicly announced their results. That chirp was exactly the sort of signal they had hoped to detect. It was the distant echo of a pair of black holes caught in the act of merging into a single, larger black hole. The physics community was electrified. It was as if we had lived for all our lives blind to the color red, and at the moment the veil was lifted we saw a rose for the first time.

And what a rose it was! Best estimates from LIGO indicated that the faint thump they recorded was the result of the coalescence more than a billion years ago of two black holes, each of them roughly thirty times the mass of the Sun. During the collision, some three solar masses' worth of energy from the black holes was vaporized into gravitational radiation.

Black holes and gravitational waves are both consequences of Einstein's general theory of relativity. General relativity predicts what sort of gravitational waves the LIGO detector should see in the event of a black hole collision, and the chirp recorded on September 14 was beautifully close to expectations. Not only is it a vindication of long-cherished theoretical ideas; this first detection event also augurs a new era of gravitational wave astronomy. The LIGO detectors saw one event of the sort we've dreamed of for decades. Now we want to explore a whole new garden of gravitational surprises.

Science seldom involves mathematical certainties, so we should ask, how sure are we that LIGO construed correctly that the little chirp is the distant echo of a billion-year-old black hole merger? Briefly, the answer is, "Pretty sure." Everything seems to fit. The signal was seen by both detectors. Nothing else seemed to be happening nearby that would explain the signal. The signal was strong enough to see with the current device, but too weak to be observed by earlier technology. The hypothesis of a black hole merger a billion years ago doesn't conflict with our general understanding of astrophysics and cosmology. The key point is that we have good hopes of seeing more such events. Indeed, LIGO announced a second confirmed event that occurred on Christmas day 2015, and a third that occurred on January 4, 2017. These events are broadly comparable to the first discovery and should give us significantly more confidence that LIGO is truly observing black hole mergers. Altogether, we believe that we are standing at the dawn of a new age in observational astrophysics—one in which black holes will play a pivotal role.

In this book, we describe black holes, both as astrophysical objects whose existence is now almost beyond doubt, and as theoretical laboratories which allow us to hone our understanding not just of gravity but also of quantum mechanics and thermal physics. Explanations of special relativity and general relativity set the stage, in Chapters 1 and 2. Then in subsequent chapters, we go on to discuss Schwarzschild black holes, rotating black holes, collisions of black holes, gravitational radiation, Hawking radiation, and information loss.

So, just what is a black hole? Essentially, it is a region of spacetime toward which matter is drawn and from which escape is impossible. Let's focus the discussion on the simplest black holes, known as Schwarzschild black holes (in honor of their discoverer, Karl Schwarzschild). There's an old saying, "What goes up must come down." Inside a Schwarzschild black hole, a stronger statement is true: Nothing can go "up." Only "down." But we are not sure where "down" eventually leads. The most straightforward hypothesis, given the mathematics behind Schwarzschild black holes, is that a terrible, infinitely compressed kernel of matter lurks at its core. Colliding with that core is the end of everything. It is even the end of time. This hypothesis is hard to test because no observer who goes into a black hole can ever report back on what he sees.

Before we go on to explore Schwarzschild black holes in more depth, let's first take a step back and consider gravity in some of its milder forms. From the surface of the Earth, if we impart a sufficiently large upward velocity to an object, then it will keep moving upward forever. The minimal velocity for which this is true is the escape velocity. Neglecting air friction, escape velocity is approximately 11.2 kilometers per second. By way of comparison, it's hard for a human to throw a ball faster than about 45 meters

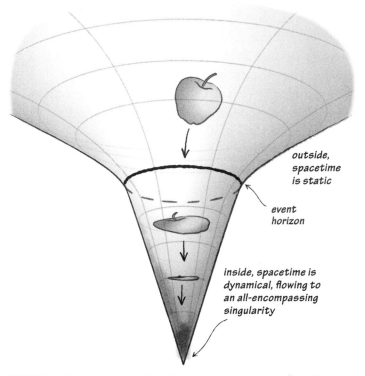

outside,
spacetime
is static

event
horizon

inside, spacetime is
dynamical, flowing to
an all-encompassing
singularity

FIGURE 0.1. Cutaway of a schematic representation of a black hole geometry. Far outside the horizon, spacetime is flat. Moving toward the horizon, it becomes ever more curved but is still time independent, or static. Crossing the horizon however, spacetime becomes dynamical: as time flows, two of the spatial dimensions (having the geometry of a sphere) compress, while the third (not shown) lengthens, until all of space is stretched and squeezed into an infinitely long and thin singularity.

per second—less than half a percent of escape velocity. The muzzle velocity of a high-powered rifle is roughly 1.2 kilometers per second—so, slightly more than 10% of escape velocity. What we usually mean, then, by "What comes up must come down" is that Earth's gravity is strong compared with how forcefully we can propel objects upward using ordinary means.

Rocketry is the modern means to conquer Earth's gravity altogether and send objects into space. To escape Earth's gravity, it is not strictly necessary for a rocket to go faster than 11.2 kilometers per second (though some rockets do). What can happen instead is that a rocket travels at a slower velocity, but it has enough fuel to keep thrusting upward until it reaches altitudes where Earth's gravitational field is significantly weaker. The escape velocity from such altitudes is correspondingly smaller. In other words, a rocket designed to take a space probe entirely out of Earth's gravitational field must be going faster than the escape velocity at the point where the rocket stops firing.

Now we could ask, what if the Earth were much denser? Escape velocity from the surface would be larger because the gravitational field would be more intense. The densest stable form of ordinary matter in the known universe occurs in neutron stars. They pack approximately one and a half times the mass of the Sun into a sphere of only 12 kilometers in radius, though this radius is not very precisely measured. Ordinary matter is completely crushed into the surface by the tremendous gravitational forces, which are something like a hundred billion times stronger than the gravitational field of the Earth. Assuming a 12 kilometer radius, the escape velocity is approximately 60% the speed of light.

But why stop there? As a thought experiment, we could imagine compressing neutron stars even further. If we compress a neutron star to a radius of about four and a half kilometers, then its escape velocity reaches the speed of light. If we go past that point, gravity changes character completely. It's no longer possible for any form of matter to hold itself up against the pull of gravity. To move forward in time means to move inward in radius. Escape is impossible. This is a black hole.

The central aim of the first few chapters of this book is to make the idea of a black hole more precise. A key concept that we'll explore is the idea of an event horizon, which is the "surface" of a black hole. It is a surface in the geometrical sense of being a two-dimensional locus in three-dimensional space. For example, in the simplest example of a Schwarzschild black hole, the horizon is a perfect sphere whose radius is called the Schwarzschild radius. The odd thing about a black hole horizon is that (at least according to conventional understanding), it is not the surface of anything in particular. At the moment you fall through it, you don't notice anything special. The only problem comes if you try to turn around and get back out. No matter how hard you try—using a rocket, a laser canon, or any other means—and no matter what help you might have from outside, it is impossible to get back outside the horizon, or even to send an SOS signal out to say you're trapped. Poetically, we might think of a black hole horizon as the lip of a waterfall, beyond which spacetime cascades ineluctably downward into a singularity that destroys all things.

Black holes are more than a thought experiment. They are believed to occur in the universe in at least two situations. One is along the lines of the previous discussion of neutron stars. When large stars run out of nuclear fuel, they collapse in on themselves. This collapse is a messy process in which a great deal of matter is blown off into the surrounding universe in an explosion called a supernova. (In fact, it's generally thought that supernovae play a crucial role in distributing metals and other moderately heavy elements throughout the universe.) Enough mass can remain behind that it is impossible for a neutron star to form and also remain stable. Instead, this remaining mass collapses to form a black hole whose mass is at least a few times the mass of the Sun. The black holes whose mergers were observed by LIGO are somewhat more massive, but still plausibly produced by stellar collapse.

Much larger black holes are thought to exist at the center of galaxies. Exactly how these black holes formed is more mysterious and may be related to dark matter, the physics of the very early universe, or both. Black holes at the center of galaxies are tremendously massive, containing thousands to billions times as much mass as the Sun. One is thought to be at the center of the Milky Way, containing some 4 million solar masses. We may well ask, how can we be sure that a black hole is present if no signal can escape from a black hole horizon? The answer is that nearby objects respond to the gravitational pull of the black hole. By tracking the motion of stars near the center of the Milky Way, we can be certain that a very massive, very dense object is present there. We cannot prove in this way that it is a black hole. What we can say is that if it isn't a black hole, then it's something much stranger. In short, black holes are the simplest possibility, and the modern consensus is that they do exist at the centers of many if not most galaxies.

Black holes are a tremendously useful theoretical laboratory because they are mathematically simple compared to most astrophysical objects. Stars are really very complicated. Nuclear reactions at the core of stars provide their power. The matter inside stars experiences pressures and fluid dynamical motions that we can simulate numerically but certainly do not understand completely. And stars have surface xiv

dynamics which are probably as complicated as the Earth's weather patterns. A black hole, by comparison, is wonderfully simple. In the absence of other matter, black holes must settle down to one of a few definite forms, all of which are explicitly understood as curved geometries which solve Einstein's equations of general relativity. To be sure, infalling matter complicates the picture, but there is a tolerably good understanding of how ordinary matter behaves as it falls into black holes. Nowadays, there is even a good numerical understanding of what happens when one black hole collides with another, and a central aim of Chapter 6 of this book is to explain how this understanding is achieved and what it means for experiments such as LIGO.

Where things get strange is that black holes aren't really black. Using quantum mechanics, Stephen Hawking showed that black holes have a definite temperature, related to their surface gravity. In fact, there is a whole field of study known as black hole thermodynamics, in which geometrical properties of black hole solutions are put into precise correspondence with properties familiar from the study of heat: temperature, energy, and entropy. There is even a proposal that black holes in distant parts of the universe may have overlapping interiors, and that these overlapping interiors help encode a quantum effect known as entanglement. We will give an introduction to these subjects in Chapter 7.

Black holes continue to grip the imagination of scientists to this day. Astronomers search for ever more precise evidence of the properties of spinning black holes, and now they eagerly anticipate collaboration with gravitational wave observatories to understand the cataclysmic events surrounding black hole mergers. And this is just the beginning for gravitational wave astronomy, with a worldwide effort under way to build a network of detectors in the United States (the two LIGO detectors in Hanford, Washington and Livingston, Louisiana), Europe (Virgo and GEO600), Japan (KAGRA) and India (LIGO India). Meanwhile, string theorists study black holes in higher dimensions not only as a means of probing quantum effects in gravity, but also as analogs of physics as diverse as heavy ion collisions, viscous fluids, and superconductors. And black holes inspire us all to ponder the strangest questions: Could black holes ever be useful to us? What really lies inside them? What would it be like to fall into one? Or—is it possible we've already fallen in and just don't know it yet?

THE LITTLE BOOK OF BLACK HOLES

CHAPTER ONE

SPECIAL RELATIVITY

To understand black holes, we have to learn some relativity. The theory of relativity is split into two parts: special and general. Albert Einstein came up with the special theory of relativity in 1905. It deals with objects moving relative to one another, and with the way an observer's experience of space and time depends on how she is moving. The central ideas of special relativity can be formulated in geometrical terms using a beautiful concept called Minkowski spacetime.

General relativity subsumes special relativity and also includes gravity. General relativity is the theory we need in order to really understand black holes. Einstein developed general relativity over a period of years, culminating in a paper in late 1915 in which he presented the so-called Einstein field equations. These equations describe how gravity distorts Minkowski spacetime into a curved spacetime geometry, for example the Schwarzschild black hole geometry that we will describe in Chapter 3. Special relativity is simpler and easier than general relativity because gravity is neglected—that is, gravity is ignored, or presumed to be too weak an effect to be significant.

Special relativity includes the formula $E = mc^2$, relating energy E, mass m, and the speed of light c. This is one of the most famous equations in all of physics, possibly in all of human understanding. $E = mc^2$ made it possible to foresee the awesome power of nuclear weapons, and it is at the core of our hopes, as yet unrealized, for a clean source of energy from nuclear fusion. $E = mc^2$ is also very relevant to black hole physics. For example, the 3 solar masses' worth of energy ejected from the first observed black hole collision is a prime illustration of the equivalence of mass and energy. To get an idea of just how cataclysmic this collision was, consider that the mass converted into energy in the explosion of a nuclear weapon (assuming a yield of 400 kilotons) is a mere 19 grams.

Special relativity is closely related to James Clerk Maxwell's theory of electromagnetism. Indeed, an early hint of the relativistic view of space and time emerged in the late 1800s in the form of so-called Lorentz transformations, which explain how observers' perceptions of electromagnetic phenomena depend on how the observers are moving. The most familiar electromagnetic phenomenon is light, which is a traveling wave of electric and magnetic fields. A consequence of Maxwell's theory is that light has a definite speed. Relativity is built around the idea that this speed is truly a constant, independent of the motion of the observer.

The motion of observers is described in special relativity in terms of frames of reference. To get an idea of what a frame of reference is, think of a high-speed train. If all

CHAPTER ONE

the passengers are seated and all the luggage is stowed, then everything on the train indeed is stationary with respect to the train itself. But the train is moving quickly relative to the Earth. Let's assume that the train is moving in a straight line at constant speed. To give a fully precise account of frames of reference, we should stipulate the absence of any significant gravitational field. For example, instead of a train running at constant speed along the Earth's surface, we would need to consider a spaceship coasting at constant speed in otherwise empty space. Earth's gravitational field is weak enough that, for present purposes, we can ignore its effects on the train and work with just the special theory of relativity rather than the general theory.

Without looking out the windows, it's hard to tell how fast the train is moving. In a situation where the train has fantastic suspension and the track is very even, and where the blinds on all the windows are down, it would be impossible to know that the train is moving at all. The train provides a frame of reference—the one that its passengers naturally use to judge whether something inside the train is moving. They can't tell (in the ideal situation just described) whether the entire train is moving. But they certainly know when someone walks up the aisle, because such a person is moving relative to their frame of reference. Furthermore, all physical phenomena, like balls dropping or tops spinning, would behave the same, as observed by an observer on the train, whether the train is actually moving or not. Briefly then, a frame of reference is a way of looking at space and time which is associated with an observer, or a group of observers, in a state of uniform motion. Uniform motion means that the train is not speeding up, or slowing down, or turning. If the train is doing one of these things, then

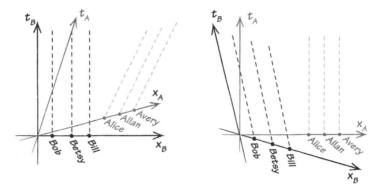

FIGURE 1.1. Left: Minkowski spacetime, showing three observers in the B-frame as stationary and three observers in the A-frame as moving forward. Right: A different perspective on Minkowski spacetime, in which the B-frame observers are now moving backward and the A-frame observers are stationary.

the passengers will notice; for example, rapid acceleration pushes them back in their seats, whereas rapid deceleration throws them forward.

Let's imagine our train passing through a station without stopping or slowing down. The passengers on the train, call them Alice, Allan, and Avery, are observers in a moving frame of reference which we'll call the A-frame. Meanwhile, their friends Bob, Betsy, and Bill stand on the platform, in a stationary frame of reference which we'll call the B-frame. To draw these frames of reference, we put B-frame position on the horizontal axis and B-frame time on the vertical axis, and we map out the trajectories of our various observers through space and time, so that over time, B-frame observers always stay at the same B-frame positions, whereas A-frame observers move forward. The resulting diagram is actually Minkowski spacetime! The word spacetime refers to the fact

CHAPTER ONE

that we are showing space and time on the same diagram. It's possible to take a different perspective on Minkowski spacetime, such that A-frame observers are shown as stationary while B-frame observers move backward. More on that perspective later.

Special relativity hinges on the assumption that the speed of light is constant. In other words, the speed of light is supposed to be the same when measured by the observers on the train as when measured by the observers on the platform. If that weren't so, then by measuring the speed of light, an observer could tell which of the two frames of reference she was in. But a core tenet in relativity theory is that physics should be the same in any frame of reference, so that you really *can't* tell which frame you're in through any physical measurement. According to this tenet, we cannot pick out a frame and say, "Remaining in this frame is what it means to be stationary. Motion consists of being in a different frame." We can only say, "Any frame is as good as any other. The only idea of motion that we can permit is motion of one observer with respect to another." In other words, states of motion are not absolute; they are relative. Thus it was a misnomer to refer to the A-frame as moving and the B-frame as stationary. All we can really say is that they are moving with respect to one another. (The idea of the B-frame being stationary seemed natural, though, because we were implicitly thinking of motion relative to the Earth.)

The intuition we've explained about relative motion seems like common sense, and we should ask ourselves how we can possibly get any leverage from it on questions relating to the deep nature of space and time. The key ingredient is Maxwell's theory of electromagnetism. What this theory tells us (among other things) is that if Alice pulls out a laser pointer and sends a pulse of light forward, toward the front of the train, and Bob does the same, then the two light pulses travel forward at exactly the same velocity. This seems like another innocuous claim, but it's not! For example, if we arrange for the train to go at 99% of the speed of light (so obviously not an American train), then wouldn't Bob measure a laser pulse shot forward by Alice to be traveling at almost double the speed of light? After all, she is moving forward at 99% of the speed of light relative to Bob, and her light pulse moves forward at the speed of light relative to her, so it seems like Bob should measure her light pulse to be moving forward at 199% of the speed of light. But according to electromagnetism, he doesn't! He measures it to be moving at precisely the same speed of light, relative to him, that Alice would report if she measured its motion relative to her.

How is this possible? The answer is that Alice and Bob measure the passage of time differently, and they also measure length differently. The details of how this happens are encoded in the Lorentz transformation, which is a mathematical expression relating time and length in the A-frame to time and length in the B-frame. A Lorentz transformation is easy to draw using Minkowski spacetime. Before the Lorentz transformation (the left side of Figure 1.1), we can think of the B-frame as stationary and the A-frame as moving forward. After the Lorentz transformation (the right side of Figure 1.1), the A-frame is stationary and the B-frame is moving backward! A Lorentz transformation is just the change of perspective between the account that Bob would offer based on thinking of his frame as stationary, and the one that Alice would offer based on thinking of her frame as stationary.

CHAPTER ONE

Key consequences of the Lorentz transformation include time dilation and length contraction. We're going to explain time dilation first because it's easier to describe. Suppose that at noon on Friday you get on a train at Princeton Junction. For convenience, we're going to say that this time and place correspond to the origin of Minkowski space, where the t and x axes cross. Now, there are fast trains and slow trains that go through Princeton Junction; some go north toward New York, and some go south toward Philadelphia; and you can decide which one you want. What you're going to do is ride the train for exactly one hour by your watch, and then get off and mark where you end up. Obviously, if you take a fast train, you get farther. But beware of the assumption that you get exactly twice as far riding a train that goes twice as fast. The tricky part is that you're riding the train for exactly one hour as measured by your own watch. The speed of a train is something observers who are stationary relative to the ground would measure, and their watches run a bit differently from yours because they're in a different frame of reference.

So where do you end up? More generally, if you and a bunch of friends all take different trains (all departing Princeton Junction at the same time), where do you all wind up? The answer is that you all wind up somewhere on a hyperbola in Minkowski spacetime (see figure 1.2). In other words, the hyperbola is the set of all possible final locations that you can reach after precisely one hour of your own travel time. One possible final location is Princeton Junction itself, at precisely 1 p.m. Princeton Junction time. The way you wind up there after an hour is if you are silly enough to spend an hour on a train that doesn't move at all. In that situation, of course it's 1 p.m. Princeton Junction time when you "arrive," because your frame of reference is the same as the

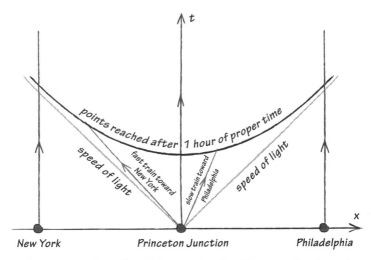

FIGURE 1.2. Trains from Princeton Junction. The curve showing points reached after 1 hour of proper time is a hyperbola.

station's, so that your watch exactly keeps pace with station time. If instead you get on a train that actually goes somewhere, your watch runs slower than station time, so when you get off after an hour of perceived travel time, station time is actually later than you think it ought to be. This later-than-you-think effect, known as time dilation, is captured in Minkowski spacetime by the way the hyperbola curves upward in the time direction as you go to locations farther and farther from your starting point.[1] Minkowski spacetime is sometimes called hyperbolic geometry, in reference to precisely the type of hyperbola we have been discussing.

1 The later-than-you-think effect for a normal train ride from Princeton to New York City amounts to approximately a hundred billionth of a second. So time dilation isn't going to make you late for work.

CHAPTER ONE

In Minkowski spacetime, we visualize the constant speed of light by drawing light rays at precisely a 45° angle relative to the vertical time axis. You'll notice that the hyperbola of possible endpoints for a one-hour train ride is wholly within the region of spacetime between two light rays emanating from the origin. This is the way Minkowski spacetime encodes the statement that none of our trains can go faster than light.

It may seem like our discussion of time dilation doesn't have much to do with Lorentz transformations. To see that it really does, let's go back to calling the reference frame of the train the A frame, while the reference frame of the Earth is the B-frame. Suppose Alice spends an hour in the A-frame on her way from Princeton Junction to New York. Meanwhile, Bob and his friends remain stationary with respect to the Earth. How should they figure out the time of Alice's arrival? It's not very useful for her to call them when she arrives, because the signal she would use could only travel at the speed of light, and Bob and friends would have to do a calculation based on the time they received her call, the speed of the signal, and the distance to New York City to figure out when Alice arrived. That all sounds too tricky. So Bob figures out a better way. He synchronizes his watch with one of his friends, let's say Bill, and Bob and Bill take up positions at the Princeton Junction and New York City train stations, respectively. Bob measures when Alice leaves, and Bill measures when she arrives. No telephony required. It might seem tricky to synchronize watches in a reliable way between distant observers, but one good strategy would be for Bob and Bill both to start out halfway between Princeton Junction and New York City, synchronize their watches while standing next to one another, and then walk at identical speeds to their respective stations, all well before Alice boards her train.

In this whole narrative of Alice's train ride, the A-frame is clearly privileged, because Alice doesn't need any friends to figure out the duration of her train ride, whereas Bob and Bill must cooperate to make their measurement of the time. The time interval that Alice measures is called proper time because she measures it while remaining at a fixed location in her own frame of reference (the A-frame). The time interval that Bob and Bill measure is dilated time, which always must be greater than proper time. Dilated time is part of how the A-frame and B-frame perspectives on spacetime are related. The Lorentz transformation between the A-frame and the B-frame contains time dilation, and more.

A similar discussion can be used to describe length contraction. Instead of a train ride, let's imagine that Bob, Bill, and Alice go to the Olympics, where Alice hopes to set a record in pole-vaulting. Her secret is that she can run really fast, at 87% of the speed of light. (For some reason, she leaves the 100 meter dash to Usain Bolt, even though she figures she could post a time of under 0.4 microseconds.) Alice chooses a 6 meter pole, which is longer than most vaulters want, but after all she is pretty exceptional. Bob and Bill don't believe that Alice is using a pole that long, so they resolve to measure it as Alice charges down the track, holding her pole perfectly horizontal as she goes. Clearly, they've got a tough job. How can they actually make the measurement? Here is what they come up with. First, they synchronize their watches. Then they stand somewhat less than 6 meters apart, and they agree that at precisely the same time, they're going to glance up at Alice and record which part of her pole they see. After many attempts, they manage to arrange themselves

CHAPTER ONE

so that Bob sees the tail end of Alice's pole, while Bill sees the front tip. Then they measure the distance between themselves. The answer is that they are only 3 meters apart. They reasonably conclude that Alice's pole is 3 meters long. They approach Alice and explain what they found. Alice protests that they can't have gotten it right. She enlists the help of her two friends, Allan and Avery, who run with her (apparently they're equally good sprinters) and measure her pole in her frame. The answer they find is that her pole is 6 meters long.

Once again, the A-frame is privileged in this discussion, because it's the frame in which Alice's pole is stationary. Its length as measured in the A-frame is called the proper length. Its length as measured in the B-frame is always shorter, and it is termed the contracted length. Time dilation and length contraction are closely linked, as we can appreciate in this example by considering what Alice would say about her experience running down the track toward the bar. As measured in her frame, it takes her half as long to get to the bar as Bob and Bill would have measured by the protocol we discussed above in reference to Alice's train ride to New York City. Time dilation, then, involves a factor of two for Alice's record-busting sprint at 87% of the speed of light. Length contraction also involves a factor of two: A-frame observers say her pole is 6 meters long, and B-frame observers say it is 3 meters long. In general, time dilation and length contraction always involve the same factor, sometimes called the Lorentz factor.

There seems to be a disconnect between our discussion of special relativity, which focuses on spacetime geometry, and the famous equation $E = mc^2$. Let's try to bridge this gap by considering a partial derivation of $E = mc^2$, where the most important steps can be illustrated geometrically. Our argument is only a partial derivation because it will involve some approximations and a couple of other formulas which we don't fully justify or derive.

The first step is to say in an equation what mass actually is. The best equation to use is $p = mv$, where p is the momentum and v is the velocity of a slowly moving massive body whose mass is m. The relation $p = mv$ comes directly from Newtonian mechanics, and it's OK for us to use it provided we make v much less than the speed of light. The next step is to relate energy to something. Here we are going to have to take yet another result from electromagnetism on faith: The momentum p of a light pulse is related to its energy E by the equation $p = E/c$. As we have learned, light pulses are peculiar in that they move at a fixed velocity in any frame. That's very unlike the way massive objects behave. In a given frame of reference, massive objects can either stand still, or they can move with some velocity v, which—according to special relativity—must always be less than the speed of light.

We now know the momentum of a massive object ($p = mv$) and the momentum of a light pulse ($p = E/c$). It would be wrong to set these two momenta equal, because massive objects are different from light pulses! What we must do instead is to figure out how to build a massive object out of light pulses. Then we will be able to use our momentum equations to derive $E = mc^2$.

Here is the crucial idea. Let's set up two perfectly reflecting mirrors exactly facing one another, and arrange for two identical light pulses to be going back and forth between the mirrors in such a way that they are always going in opposite directions. We claim that this setup is, effectively, a massive body. Assume that we can make the mirrors very light—so

CHAPTER ONE

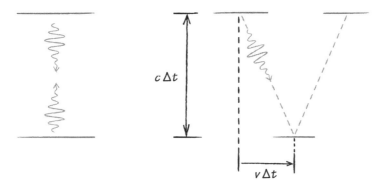

FIGURE 1.3. Left: Two identical light pulses travel back and forth between two mirrors. Right: The mirrors move to the right with a velocity v. In the time Δt that it takes one of the light pulses to get from one mirror to the other, the light pulse travels a distance of approximately $c\Delta t$ upward or downward, and a distance $v\Delta t$ sideways.

light that we can ignore them in calculations of the mass as well as the energy. Then the energy of our massive body is twice the energy of one of the light pulses. Its momentum is exactly zero because one light pulse has upward momentum at the moment when the other has downward momentum, and these upward and downward momenta cancel. They cancel because the body as a whole has no upward or downward motion; only its parts are moving.

To arrive at the right argument leading to $E = mc^2$, what we need to do is coax our whole light-and-mirrors contraption into motion. We're going to simplify the discussion by tracking the behavior of only one light pulse. If we tracked them both, we'd just get double the energy and double the mass. It's also going to simplify our discussion for the motion of our contraption to be sideways relative to the original up-and-down motion of the light pulse that we're tracking. Once this motion is in progress, the light pulse isn't just going up and down anymore. It's going a little sideways too. This is where geometry starts to come in. The light pulse's sideways motion is at a speed v, while its up-and-down motion is at a speed c. (Actually, its up-and-down motion is just a little slower than c because the *total* velocity of the light pulse is c. At the accuracy we need, it's OK to ignore this detail.) Another way to put it is that a fraction v/c of the motion of the light pulse is sideways. So it seems reasonable to assert that the sideways momentum $p_{sideways}$ of the photon is v/c times its total momentum $p = E/c$. That is, $p_{sideways} = Ev/c^2$. We now assert that $p_{sideways} = mv$, which makes sense because $p_{sideways}$ is the sideways momentum of the contraption as a whole (tracking only one of the two light pulses), and we're thinking of the contraption as a massive body. If we now combine our two ways of writing $p_{sideways}$, we get $\frac{Ev}{c^2} = mv$. Simplifying this equation, we get . . . drum roll . . . $E = mc^2$!

It might be objected that our light-and-mirrors contraption is very unlike the massive objects of everyday experience. That's not quite true. Protons and neutrons comprise most of the mass of everyday materials, and they can be approximately understood as little regions of spacetime inside which three nearly massless quarks bounce around at close to the speed of light. If this were the whole story, then the mass of the proton would come entirely from the motion of its constituent quarks, just as the mass of the light-and-mirrors contraption comes from the light pulses. In fact, there is more to the story: Quarks interact strongly with one another, and these interactions also contribute significantly to the total energy of the proton and hence its total mass. Nevertheless, the essential origin of most of the mass

CHAPTER ONE

of everyday matter actually has more to do with our light-and-mirrors analysis than any intrinsic mass of fundamental constituents of matter.

The further we go with special relativity, the clearer it is that Maxwell's theory of electromagnetism is a key precursor to it. Better yet, it is in many ways a precursor to general relativity! Let's end this chapter with a tour of the highlights of Maxwell's amazing theory.

Before electromagnetism was properly developed, people understood the attraction between positive and negative charges in much the same way that Newton understood the gravitational pull between the Earth and the Sun. Briefly, they didn't really understand either one. Newton knew he didn't understand. He wrote of his quest to understand the origin of gravitational pull, "I have not as yet been able to discover empirically the reason for these properties of gravity, and I frame [or feign] no hypothesis." (This is an approximate translation from Newton's original Latin.) Of course, Newton had a highly useful quantitative law describing the strength of the gravitational pull. In particular, he knew that the pull weakens as the inverse square of the separation between the gravitating bodies. The attractive force between positive and negative charges follows a similar inverse square pattern. But it bothered him and his many successors that there could be *any* force acting over a distance. In other words, it's strange that a force on one object can be caused by the existence of another object which is far away. Michael Faraday championed the modern resolution of this puzzle. According to his ideas, a charged object both creates and responds to electric fields, which spread out in space according to four equations whose final form Maxwell discovered.

SPECIAL RELATIVITY

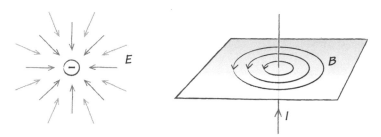

FIGURE 1.4. Left: The electric field E near a negative charge points everywhere inward. Right: A wire carrying a current I creates a magnetic field B that circulates around it.

In Faraday's picture, negative charges do *not* directly attract positive charges. Instead, a negative charge orients the nearby electric field so that it points directly toward the negative charge. The electric field in turn pulls on a positive charge at some distance away from the negative charge, and the net result is that the positive charge is drawn toward the negative charge. We could equally well say that the positive charge orients the nearby electric field to point directly away from it, and the electric field in turn exerts a pull on the negative charge. Both effects are happening at once. If all we watch is the charges, we conclude (correctly) that they feel equal and opposite forces drawing them together. Faraday's point is that these forces arise only through the action of the electric field, which has an existence independent of whatever charges might have produced it.

A similar story can be presented for magnetic forces and magnetic fields. Without entering into details, moving electric charges both create and respond to magnetic fields, which spread out in space in a fashion dictated by Maxwell's equations. A particularly important example is that magnetic fields form around a wire carrying an electric current. Electric current is the motion of microscopic charges inside the wire, so this is just a special case of the more general rule that moving charges produce magnetic fields.

Like electric fields, magnetic fields are supposed to have some existence independent of any particular configuration of moving charges that might produce them. To understand what we mean by this, let's consider a setup used by Maxwell in his development of the final form of electromagnetism. Put two metal plates parallel to one another without touching, and attach a wire to each one. This setup is known as a capacitor. Let an electrical current flow into one plate and out of the other. This flow results in an increase of positive charge in one plate over time (actually, a growing deficiency of electrons) and an equal increase of negative charge in the other plate (a superabundance of electrons). Because of the growing charge imbalance in the plates, there is a growing electric field between the plates. That electric field runs from the positively charged plate to the negatively charged plate, and as the charges of the plates grow in magnitude, so does the magnitude of the electric field.

We know that a magnetic field forms around a current-carrying wire. In particular, a magnetic field forms around the wires supplying current to the capacitor. But there is no current flowing from one plate to the other, and naively it would therefore seem that no magnetic field should be expected between the plates. Maxwell found this hard to reconcile with his understanding of capacitors, and he proposed an amazing solution: An increasing electric field generates a circulating magnetic field in the same way that a current does. This idea is an important step beyond the

SPECIAL RELATIVITY

original notion that charges produce and respond to fields, because now we see that fields produce fields.

Actually, it was previously understood (by Faraday) that an increasing magnetic field generates a circulating electric field; this is essentially the principle on which electrical generators work. Two of Maxwell's four equations basically formalize these two reciprocal relations between electric and magnetic fields. The other two equations are simpler, saying that magnetic fields have no sources or sinks, whereas the only sources or sinks of electric field are positive and negative electric charges. All of Maxwell's equations are differential equations, which means that they are framed in terms of the rate of change of electric and magnetic fields over time, as well as the way in which these fields vary over space. The differential equations depend on the way fields behave in very small neighborhoods of spacetime. There is no action at a distance in Maxwell's equations. Everything is framed locally in terms of how nearby fields push and pull on one another.

Maxwell's greatest triumph was to show that his equations imply the existence of light. Light, as Maxwell understood it, is a combination of fluctuating electric and magnetic fields, where the spatial variation of the electric field causes the time variation of the magnetic field, and vice versa. The physical constants entering into Maxwell's equations describe the strength of electrostatic and magnetic interactions, but when they are combined in the right way, they give a numerical prediction for the speed of light—a prediction that can be verified experimentally.

Looking ahead, we will want to understand two crucial connections between electromagnetism and general relativity: Both theories involve Faraday's field concept, and both culminate in differential equations for the behavior of fields

CHAPTER ONE

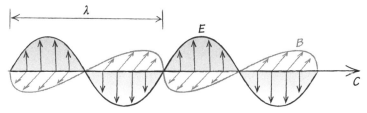

FIGURE 1.5. A light ray is a traveling disturbance of electric fields E and magnetic fields B, all moving in the same direction at the speed of light c. Interpreted as drawn to scale in the print edition of this book, the wavelength λ shown here is several centimeters, which is in the microwave range, a bit shorter than wavelengths found inside a typical microwave oven.

which imply some form of radiation. In electromagnetic radiation, electric fields beget magnetic fields, and vice versa, in a self-sustaining cascade through spacetime described by Maxwell's equations. This cascade has a characteristic wavelength, across which electric fields and magnetic fields vary from zero to their maximum value, back through zero to another maximum, and once again back to zero. Visible light is a special case in which the wavelength is about half a micron. Increasingly longer wavelengths lead to infrared light, microwaves, and radio waves, while shorter wavelengths give rise to ultraviolet light, X-rays, and gamma rays.

Einstein found the gravitational analogs of Maxwell's equations, and they are the main content of the general theory of relativity. The fields in Einstein's equations are stranger than electric and magnetic fields: Surprisingly, they are the curvature of spacetime itself. Another big surprise is that general relativity can describe massive objects in terms of pure geometry. This is quite unlike electromagnetism, in which charges remain fundamental throughout. These purely geometrical massive objects are black holes.

SPECIAL RELATIVITY

CHAPTER TWO

●

GENERAL RELATIVITY

IN SPECIAL RELATIVITY, SPACETIME IS AN EMPTY STAGE. Observers and light rays move across it, and we can talk sensibly about the time between two events or the distance between two objects, provided we keep in mind ideas like proper time, proper length, time dilation, and length contraction. The core belief that all motion is relative highlights how empty spacetime really is. If there were something "there"—some sort of stationary "ether" filling all of spacetime, then we could develop a notion of absolute motion by working always in the ether's own reference frame and describing objects as stationary or in motion with respect to the stationary ether.[1]

---

1 The ether may seem fanciful, but historically it was taken very seriously. In fact, the design of LIGO detectors is closely related to the so-called Michelson interferometer, whose original use in the late nineteenth century was precisely to measure the speed of light in different directions so as to discern the Earth's motion relative to the ether.

General relativity has quite a different feel to it. Space-time is the primary actor. It curves in response to massive bodies in a manner controlled by Einstein's field equations, which take the form $G_{\mu\nu} = 8\pi G_N T_{\mu\nu}/c^4$. Let's have a first look at what the symbols in this equation mean. The subscripted Greek letters μ and ν are the hallmarks of so-called tensor notation, which enables us to write the ten separate field equations all at once. The Einstein tensor $G_{\mu\nu}$ is a description of spacetime curvature. The stress-energy tensor $T_{\mu\nu}$ describes the presence of matter. In empty space, we would set $T_{\mu\nu} = 0$. Newton's constant G_N describes how strongly matter influences spacetime. As always, c is the speed of light. The factor of 8π, where $\pi = 3.14159\ldots$, is a relatively unimportant constant. We could redefine G_N to include the 8π, but we don't because G_N also enters into Newton's description of gravitation, so it's too late to change its meaning now.

As students of relativity, we may well ask, how can it be that general relativity subsumes special relativity while assigning such an active role to spacetime? The answer is that, in most circumstances, gravity is really weak. If we ignore gravity altogether, then we can go back to considering Minkowski spacetime, which has no curvature and which encodes most of what makes special relativity work the way it does. In particular, Minkowski spacetime is the same before and after a Lorentz transformation, which is the mathematical way of saying that all reference frames are equivalent. In the presence of gravity, the equivalence of reference frames is lost (at least in the usual sense of special relativity) because the gravitating body makes one reference frame special. The reader may recall that we tripped over this very point in Chapter 2 when we first described Bob's

B-frame as stationary, when in fact it was only stationary *with respect to the Earth*.

Even when gravity is present, we can often get away with using special relativity in small regions of spacetime. This is because weak gravity curves spacetime just a little, and if we focus on objects and events which are close enough together in time and space, we can describe them to an excellent approximation as if spacetime were flat. For example, consider shooting a bullet through an apple just as it falls from the tree. We admit that gravity is operating, and if we gave it enough time, it would cause the apple to fall downward with some measurable velocity. But in the short time it takes for the bullet to pass through the apple, gravitational acceleration is so weak that it doesn't significantly enter into the story. If we were to ask questions about the proper and dilated time elapsed while the bullet passes through the apple, special relativity would be enough.

To get an idea of how different things are when gravity *is* important, imagine shooting a bullet through a black hole. It wouldn't work! Once the bullet passes through the horizon, it's gone, and no part of it will come out the other side. This is not so much because black holes are big; it would be true even for a black hole whose horizon is the size of an apple. Black holes are such highly curved regions of spacetime that they hijack the future of every object that falls into them. (By the way, a black hole whose horizon is the size of an apple would have a mass approximately five times the mass of the Earth.)

Initially, we're going to hone our intuition about general relativity by considering gravity in situations where it is fairly weak, like the gravity we experience here on Earth. There are still some pretty strange concepts to get used to,

CHAPTER TWO

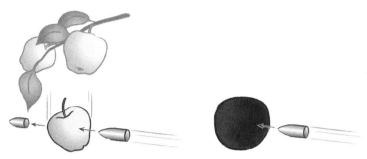

FIGURE 2.1. Left: A bullet passes through an apple just as it starts falling from the branch. Special relativity is sufficient to describe this situation because gravity is so weak and has so little time to act that it doesn't make a difference. Right: A bullet is shot into a black hole whose horizon is the same size as the apple. The bullet never comes out the other side!

most notably a new way in which time runs faster or slower depending on your position in a gravitational well. At the end of the chapter, we will turn to the powerful language of differential geometry as we revisit the Einstein equations in their full glory. Only through this language can we fully articulate the ideas of subsequent chapters—in particular, the curved spacetime geometry that is a black hole.

As much as possible, we want to take our cues on how to develop general relativity from analogies with electromagnetism. Thus, we should somehow start with the field concept and wind up with field equations that imply radiation. Our eventual goal, the Einstein field equations, are differential equations framed locally in terms of how nearby bits of curved spacetime push and pull on one another. But addressing the full, complicated story of strongly curved spacetime is exactly what we don't want to do right away, and that's why we restrict our attention for now to what we'll call "ordinary gravity." What we mean by this is gravity in situations where all massive bodies of interest are moving relative to one another much more slowly than the speed of light, and none of them are nearly dense enough to form a black hole. The solar system is such a place, and so is most of the rest of our galaxy, except near collapsed stars and black holes such as the one lurking at the galaxy's center. In discussing ordinary gravity, we are restricting attention to situations where spacetime is almost but not quite flat.

The simplest manifestation of the field concept in electromagnetism is the electric field that mediates the attraction of positive and negative charges. Our first step toward general relativity is to learn how ordinary gravity can be explained in terms of a property like the electric field—a property that is meaningful everywhere in spacetime, whether or not gravitating bodies are nearby. We are, in short, trying to find the answer that eluded Newton when he wrote of gravity's origin, "I frame no hypothesis."

The answer is time itself. More precisely, ordinary gravity arises because of gravitational redshift, which is the way time runs slower when you are close to a massive body. Gravitational redshift was first observed directly in 1959 by Robert Pound and Glen Rebka in an experiment which we will soon describe. Gravitational redshift is subtle (about one part in a billion on the Earth's surface), but it is large enough to be an important effect in the design of global positioning system (GPS) satellites. The satellites are significantly higher up in the Earth's gravitational well than we are at the surface, and as a result their clocks run faster than ours do. Precision timing is crucial to the ability of the GPS system to determine locations with high accuracy, and relativistic effects have been built carefully into the system.

CHAPTER TWO

The flow of time is also crucial to the understanding of black holes. As we'll explore in more detail in Chapter 3, spacetime near a black hole becomes so distorted that time as normally defined comes to a complete stop when you reach the horizon. As we describe the detailed properties of gravitational redshift, it's good to keep in mind that all the points we make carry over to black hole spacetimes, provided we don't venture too close to the horizon. In Chapter 3, we will complete our account of a black hole by venturing ever farther into its gravitational well until we are at last destroyed by the singularity at its core.

The idea of time slowing down near a massive body seems pretty slippery. How would we know that it's happening? And why in the world does it cause a gravitational pull on other massive bodies? The Pound-Rebka experiment gives a beautiful answer to the first question. Answering the second question will lead us eventually to the crucial notion of a spacetime geodesic.

Pound and Rebka measured gravitational redshift using—what else?—a light pulse. They knew from their study of radioactive isotopes that iron-57 (a form of iron with 26 protons and 31 neutrons) can absorb and emit a photon with an extremely well-defined frequency, which is approximately 3 billion billion hertz. By way of comparison, New Jersey 101.5 broadcasts at a much lower frequency of just over 100 million hertz. A hertz is 1 oscillation per second, and a million hertz means 1 million oscillations per second. For our purposes, it is sufficient to think of iron-57 as a tiny clock which ticks 3 billion billion times a second. These "ticks" can be observed at a distance because the photon that iron-57 emits carries them from one place to another. Pound and Rebka sent photons from iron-57 from the bottom to the top

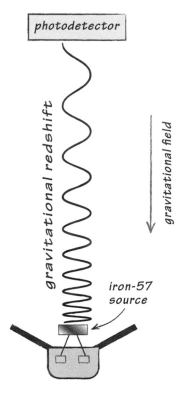

FIGURE 2.2. A schematic depiction of the Pound-Rebka experiment. Photons from iron-57 are sent upward, against the pull of gravity. A photodetector at higher elevation measures them gravitationally redshifted. Longer wavelength means redder light. The redshift in the actual experiment was far more subtle than what we have depicted in the figure.

of a tower, a bit more than 22 meters upward. They had a way to measure the frequency of those photons at the top of the tower with extraordinary accuracy. Their measurement method was analogous to the way you tune a radio to receive New Jersey 101.5 instead of some other radio station at a different frequency. They found that the frequency was smaller at the top of the tower than at the bottom. This decrease in frequency is exactly what gravitational redshift predicts.

From the Pound–Rebka experiment we can already get a glimmer of why gravitational redshift has something to do with gravitational pull. To do so, we need another of Einstein's insights (in this case following up on ideas of Max Planck): The energy of a photon is proportional to its frequency. So when the frequency decreases, the energy decreases. It actually makes sense that the energy of a photon decreases as it travels upward, because to go up, it has to fight the pull of gravity. It can't lose energy by slowing down—in general relativity just as in special relativity, light always travels at the same speed. Instead, its energy loss is encoded in its gravitationally redshifted frequency.

Having learned about time dilation in Chapter 1, you might suspect that gravitational redshift occurs because things that fall down into a gravitational well build up some appreciable speed in the process, and the resulting time dilation is what we're talking about when we say gravitational redshift. Not so. Gravitational redshift is something new and different. Pound and Rebka's clocks were both stationary with respect to the Earth.

Gravitational redshift is all around us. For example, your head ages more quickly than your feet due to gravitational redshift—assuming you're not lying flat. Like time dilation, the effect is numerically very small in ordinary experience: For example, over the course of your lifetime, your head ages a few tenths of a microsecond more than your feet. To get a more pronounced effect, you would have to stand in a far more intense gravitational field than the Earth provides. For example, if you could stand with your feet just a few centimeters above a black hole horizon whose circumference is the same as the Earth's, then your feet would age quite a bit more slowly than your head, which remains well over a meter above the horizon. Of course, it would be an absolutely crushing experience to

GENERAL RELATIVITY

stand in such an environment. We're discussing conceptual possibilities only.

How do we account for more ordinary gravitational phenomena, like falling apples and orbiting planets, using the idea that gravity's root cause is the way time runs slower close to massive bodies? We need to take a detour into ideas popularized satirically by Voltaire in the Panglossian motto: "All is for the best in this best of all possible worlds." The scientists and mathematicians of Voltaire's day, most notably Joseph-Louis Lagrange, became convinced that the motions of massive bodies—such as falling apples and orbiting planets—are in some sense optimal. In other words, the smoothly accelerating downward plunge of an apple from the tree to the ground is in some sense better than any other motion between the same initial and final states. Lagrange's great accomplishment was to formulate this idea in precise mathematical terms. In his description, any conceivable motion of the apple between prescribed initial and final states is assigned a so-called action. The actual motion that the apple chooses is supposed to be the one that either minimizes or maximizes the action. In any case, the actual motion is the best one possible in a mathematically well-defined sense.

To a disciple of Newton, Lagrange's formulation of mechanics as an optimization problem might seem like nonsense. How could an inanimate object choose the optimal path among many possibilities? The way it's supposed to work, according to Newton, is that things move in straight lines until a force pushes on them, and then they change direction according to $F = ma$. What's optimal about that? The magic is that by constructing the "action" of a moving body very carefully, Lagrange was able to recover precisely Newton's laws, nothing more, nothing less. His choice of action was, admittedly, a little nonintuitive. But if we fast-forward to general relativity, the full value of Lagrange's formulation becomes apparent. The action of an object is just the time elapsed for an observer who moves with the object. The motion that an object actually executes optimizes the proper time that elapses for that object. This is the principle of optimal proper time. In cases we will consider, proper time is maximized.

An example from special relativity helps focus the discussion. (Remember, special relativity means that we are not yet including gravity.) It's called the twin paradox. Here's how it works. Two observers, whom as usual we name Alice and Bob, start out together with identical stopwatches. We give Alice a spaceship, and the plan for her is to fly away from Bob for one day at a constant speed (let's say half the speed of light to be definite); then she turns around and returns to Bob. Meanwhile, Bob stays where he is and does nothing. If we recall our discussion of proper time from Chapter 1, we can anticipate the result of the experiment: The duration of Alice's trip as measured by Bob, using his stopwatch, will be more than the two days that Alice measures using her stopwatch. In fact, with the numbers given, Bob will measure the trip to have lasted approximately 2.3 days.

The twin paradox comes from the following faulty reasoning. All motion is relative. According to Alice, it is Bob who moves away from her, and then comes back. Shouldn't she expect that he will measure the lesser time for the trip?

To see that this reasoning is faulty, we need to identify a clear difference between Alice and Bob. The difference is that Alice accelerated when she turned around to come back. Bob never accelerated. For instance, we could let him float freely in empty space for the whole duration of Alice's trip. From the Lagrangian point of view, what Bob did was "best," because it was absolutely natural, requiring no external agency. So it makes sense that he experienced a larger elapsed proper time.

There is a fascinating variant of the twin paradox which brings gravitational effects into the picture (see figure 2.3). Let's say Alice and Bob live deep in a gravitational well, where they both go to school. They have a difficult homework assignment which is due in 48 hours, say at 9 a.m. on Monday. Bob concludes from his experience with the twin paradox that he will have the most time to work on the assignment if he moves as little as possible. So he proceeds at a very slow, steady pace to school, arriving there at 9 a.m. on Monday, working on his homework all the while. Alice, ever the adventurer, figures that she should get in her rocket ship and get up out of the gravitational well, because then the absence of gravitational redshift will give her more time to do her homework. But she is worried that the time dilation that she experiences while traveling up and down might be more important.

The principle of optimal proper time says that to maximize her proper time, Alice should do whatever inert matter would do under the same circumstances. What does matter naturally do? OK, we admit that it likes to sit still. So it sounds like Bob's plan of minimizing his motion by proceeding very slowly to school is best. But gravity changes things. Matter in a gravitational well really *doesn't* like to sit still. It likes to fall. It's unnatural for Bob to walk slowly to school in the presence of gravity; he can do it only because he's walking on top of a pile of other matter that's even deeper in the gravitational well than he is. If we want an inert hunk of matter to start at Bob and Alice's home at 9 a.m. on Saturday and wind up at their school 48 hours later, we should throw it so that it travels in a high-arcing trajectory, landing at their school at 9 a.m. on Monday. Having figured this out, Alice happily gets in her rocket, blasts off with a huge kick from her rocket thrusters, and then coasts for the rest of the weekend while frantically working on her homework.[2] Her rocket is now acting as a ballistic missile, which means that except for the initial kick, it is moving under the influence of gravity and no other forces. In other words, it's in free fall.

Alice and Bob's experiments with elapsed time help illustrate Einstein's principle of equivalence. In its simplest form, the principle of equivalence says that acceleration is indistinguishable from gravity. The key feature of the original, nongravitational form of the twin paradox is that Alice must accelerate to turn around and come back toward Bob. If we make that acceleration slow and steady instead of abrupt, then it's equivalent to having Alice spend the whole trip in a gravitational field. In contrast, the key feature of the gravitational variant of the twin paradox is that Alice spends her weekend in free fall, whereas Bob spends his in a gravitational field. Thus we see that Alice and Bob effectively exchange roles between the two versions of the paradox.

A more everyday example of the principle of equivalence is that when riding an elevator, we feel heavier when the elevator accelerates upward and lighter when it accelerates

2 You may be worried that the initial acceleration could have a large effect on the proper time. In fact, the principle of optimal proper time tells us to compare trajectories with the same initial and final *positions,* but possibly different initial *velocities.* To give a fully precise account of the twin paradox and its gravitational variant, we should allow Alice to have some initial velocity at the moment that we start her clock. Likewise, she has some final velocity when she returns to Bob's position, and we should stop the clock at the moment she reaches him so that we don't have to worry about how she brakes to a halt.

GENERAL RELATIVITY

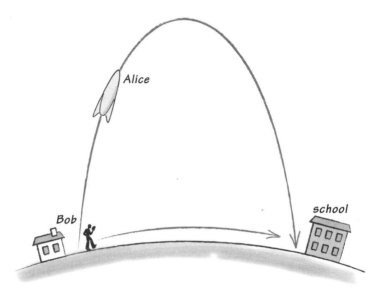

FIGURE 2.3. Bob works on his homework while walking slowly to school. Alice blasts off in a rocket and completes her assignment while flying. If Alice's rocket accelerates all at once and then glides for the rest of its motion, then she will have more time than Bob to do her homework before the school bell rings on Monday morning.

downward. If the elevator accelerated upward through empty space, with no gravitating bodies nearby, our experience as observers inside the elevator would be identical to what we experience when the elevator remains stationary in Earth's gravitational field. Likewise, if an elevator is in free fall in Earth's gravitational field, we would feel the same weightlessness inside it that we would if we were floating freely in empty space.

To move back toward Einstein's equations, let's be bold and give the rate of flow of time its proper mathematical name: It is the lapse function. In other words, the lapse

CHAPTER TWO

function tells us the rate at which time is elapsing at any given point in space. A differential equation similar to one of Maxwell's equations provides a rule for calculating the lapse function in the presence of an arbitrary collection of slow-moving masses. Knowing the lapse function, one can then turn to the principle of optimal proper time to ascertain the trajectory of a massive body responding to the gravitational field.

The differential equation for calculating the lapse function in the presence of slow-moving masses is actually a special case of one of Einstein's equations. There are nine more functions, similar to the lapse function, that fully specify the shape of curved spacetime, and loosely speaking, there is an Einstein field equation for each one of them. All together, these ten functions comprise what's called a spacetime metric. The spacetime metric is a rule for figuring out the distance between nearby points as well as the rate of time flow. Once we start talking about metrics, we have truly entered the territory of differential geometry, which is the study of arbitrarily curved surfaces, as well as higher-dimensional curved geometries, including the curved spacetime geometries of general relativity.

Our discussion of "ordinary" gravity may leave you with the idea that space remains perfectly flat while time runs at different rates in different places. That's not quite true. Actually, space opens up a bit in regions where time runs slower. To understand what this means, imagine enclosing the Earth in a perfect sphere whose area you measure carefully. Next, you measure the radius of the sphere. (Admittedly, this would require drilling a hole to the center of the Earth, but we're assuming we have appropriate superpowers.) Ordinarily, you would find that the area A and the radius r are related by the formula $A = 4\pi r^2$. But due to the presence of the Earth, r is a little bit bigger relative to A than agreement with $A = 4\pi r^2$ would require. Another way to say it is that the volume inside the sphere enclosing the Earth is a bit bigger than the volume inside a sphere of identical surface area that encloses empty space. Like gravitational redshift, the expansion of space near massive bodies is a subtle effect as long as we restrict our attention to ordinary weak field gravity. In fact, it turns out that spatial lengths (suitably defined) expand by just about the same amount that time slows down. It may now seem like all our previous discussion of falling bodies was misleading, because we were assuming that gravitational redshift is the only effect of gravity. What saves the day is that observers who move slowly relative to gravitating bodies are much more sensitive to the slowing down of time than to the expansion of space. We do have to assume that we're dealing with ordinary gravity, which requires in particular that no gravitating body of interest is nearly dense enough to form a black hole. As we leave this simplifying assumption behind, we have to delve deeper into differential geometry to understand what's going on.

Differential geometry (at least, the part we need) is centered on three concepts: metrics, geodesics, and curvature. All of them can be illustrated by considering curved surfaces, like the surface of the Earth. The metric is easy, because it's all about distance—or, at least, it seems easy at first. It's approximately 2,440 miles from Washington, DC to San Francisco as the crow flies. What we really mean by that is if you travel along (or just above) the Earth's surface, the shortest path from Washington, DC to San Francisco is 2,440 miles. As points in space, though, the two cities are a little

CHAPTER TWO

bit closer, more like 2,400 miles apart. The slight difference comes from the fact that if we could go straight through the Earth, we would cut off a little distance compared to traveling on the surface. When we travel along the surface, our path is necessarily curved; to find the total distance, the natural approach is to divide up the path into little bits, each of which is almost straight, and then add up the lengths of all the little bits. The term "differential" refers to this process of chopping things up and measuring the little bits. The task of a metric in differential geometry is to tell us the lengths of the little bits. When we want to know the total length of a path, it's assumed in differential geometry that we're up to the task of adding up the lengths of all the little bits, which is ultimately an exercise in integral calculus.

A geodesic along the surface of the Earth between Washington, DC and San Francisco is the shortest path accessible to a traveler who stays on the surface. A geodesic is not a straight line, but it is as straight as any path along the Earth's surface can be. By "straight," what we mean is that to travel along a geodesic on the Earth's surface from Washington, DC to San Francisco, we would go straight and avoid making any turns. Because of the curvature of the Earth, that straightest possible path goes a little farther north in latitude than the location of either city. A more pronounced example of this phenomenon is that airplanes take a polar route on long trips—say, from Athens to San Francisco. It turns out that the shortest path between those two cities takes you over Greenland, much higher in latitude than either city. (Of course, airplanes travel above the Earth's surface rather than on it; but compared to the Earth's radius their cruising altitude is negligible, so for present purposes we can think of airplanes as traveling on the surface of the Earth.)

GENERAL RELATIVITY

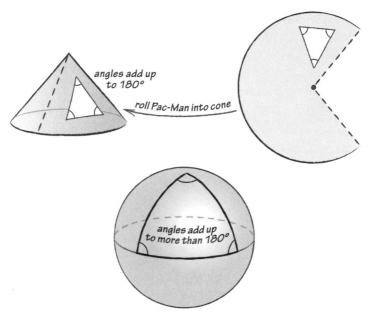

FIGURE 2.4. A cone has no intrinsic curvature, because it can be produced by rolling up a Pac-Man shape. As a result, when a triangle is drawn with sides which are segments of geodesics, the angles add up to 180°. The same triangle drawn on the Pac-Man shape before rolling it up has sides which are ordinary straight line segments. In contrast, a sphere has positive intrinsic curvature, so a triangle with geodesic sides has angles that add up to more than 180°.

Curvature seems like it should be easy at first, since we are all familiar with the way the Earth's surface curves. But there is actually something very subtle about the notion of curvature most often used in differential geometry (and needed for relativity). To appreciate the subtlety, consider the difference between a cone and a sphere. Both are curved, but they are curved in very different ways. A flat piece of paper can be rolled into a cone without stretching it, but if you try to cover a sphere with a flat piece of paper, you will have to crease or tear the paper. We therefore say that a sphere is "intrinsically curved," whereas a cone is "intrinsically flat" (except right at the tip). Both a sphere and a cone have "extrinsic curvature," which refers to the usual idea that they are curved as surfaces in three dimensions. In relativity, intrinsic curvature is all that matters. To focus on intrinsic curvature in the discussion of curved surfaces, we restrict ourselves to questions that can be answered by measurements performed on the surface only. Taking this attitude, we would say that the distance from Washington, DC to San Francisco is 2,440 miles, and we're not even going to think about the shorter path between them that runs through the Earth.

To get a better handle on what intrinsically curved geometry means, we should think about triangles whose legs are geodesics. In a flat two-dimensional geometry, the angles at the corners of any such triangle will add up to 180°. In the presence of positive intrinsic curvature, like the curvature of the Earth's surface, the angles will add up to more than 180°. It turns out there are curved surfaces (shaped somewhat like the neck of an hourglass) on which the angles at the corners of triangles whose legs are geodesics add up to less than 180°. This is negative intrinsic curvature.

Now that we have introduced the main ideas of differential geometry, let's see how they generalize to the four-dimensional spacetimes of general relativity.

The metrics used in general relativity are a little more complicated than the metric on the surface of the Earth, because they accomplish two distinct tasks. One is to determine the distance between two spatially separated events, and the other is to determine the time elapsed between two

temporally separated events. The time between two temporally separated events is precisely the time that would elapse
for a freely falling observer between when she observes one
event and the other—assuming that both events take place
at the same location in her frame of reference. Spatially separated events are more difficult to conceptualize, because by
definition they are events separated so widely in space that
no observer moving at less than the speed of light can observe them both at the same location in his frame of reference. For a static spacetime (one which does not change over
time), we can define the distance between spatially separated events in terms of how long it takes a signal from one
to be received at the location of the other. Metrics are of
absolutely central importance to general relativity, in that
solutions to Einstein's equations are spacetime metrics. Our
discussion of black holes in Chapters 3 and 4 will hinge
on particular spacetime metrics known as the Schwarzschild
and Kerr solutions.

As we mentioned previously, a metric in general relativity
comprises ten functions, one of which is essentially the lapse
function, which tells us how fast time runs. Another of these
ten functions is the one that tells us how space opens up in
the presence of massive bodies. The other eight functions
describe various distortions of spacetime, a little bit similar
to amusement park mirrors which show you an image that is
stretched in one direction or another. We can pack all these
ten functions into the so-called metric tensor, usually written $g_{\mu\nu}$—not to be confused with the Einstein tensor $G_{\mu\nu}$!

Geodesics in relativity are also a little more complicated
than on curved surfaces, partly because they come in three
varieties. A spacelike geodesic is the shortest path between
two spatially separated points, similar to a straight road along

CHAPTER TWO

the surface of the Earth between Washington, DC and San Francisco. But a spacelike geodesic is a path which no observer can follow, because to do so would mean exceeding the speed of light. This may seem like nonsense—how can it be impossible to follow the shortest path between two points? The issue is that a geodesic in spacetime specifies not only where you are supposed to go, but also when you are supposed to get there. A good example of a spacelike geodesic is a line segment at constant time between two points in Minkowski space. "Following" this geodesic would mean that you arrive at precisely the same moment you leave, and of course that's impossible.

The second type of geodesic is a timelike geodesic, which is the type of path that massive bodies naturally follow if acted on by no forces other than gravity. Alice's ballistic motion in a gravitational field is one example of a timelike geodesic, and Bob's floating freely in place, away from any gravitational field, is another. Timelike geodesics maximize proper time, as we discussed in connection with our several versions of the twin paradox. Indeed, the principle of optimal proper time finds its full expression in the requirement that massive bodies in arbitrary curved spacetimes should follow timelike geodesics.

There is one more type of geodesic in general relativity, namely a null geodesic. It is the natural path of a light ray in curved spacetime. We sometimes prefer to call the geodesics in general relativity "spacetime geodesics," to emphasize that they have information about time as well as space. But most practitioners just say "geodesic," and we'll adopt that abbreviated terminology from now on.

When we pass from two-dimensional surfaces to four-dimensional spacetimes, curvature becomes quantitatively

40

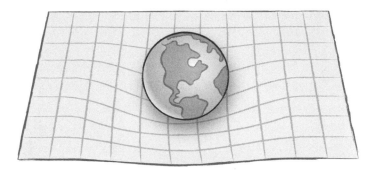

FIGURE 2.5. The Earth causes a bulge in space, which is often depicted by showing space sagging downward. Space does curve near a massive body, but the curvature is intrinsic, having to do with distortions of space within itself rather than some kind of bending of space into an extra dimension.

more complicated, but conceptually it's similar: When we ask questions about the angles at which geodesics meet, the answers can be different from our flat space expectations in ways captured by the so-called Riemann curvature tensor. The Einstein tensor $G_{\mu\nu}$ is a stripped-down version of the Riemann curvature tensor which picks out just those aspects of spacetime curvature that are affected by the presence of mass (or energy, or momentum, or pressure, or shear stress).

At least according to current understanding, there is nothing beyond the four dimensions of spacetime for it to curve into. Good questions about curvature in general relativity are ones that can be answered referring to geodesics in four-dimensional spacetime. We don't ever need to think of "short-cuts" through some larger ambient geometry, like the shortcut through the Earth from Washington, DC to San Francisco. Admittedly, when we draw curved spacetime in figures illustrating the effects of gravity, we depict it as a two-dimensional membrane which sags downward in the presence of massive bodies. That depiction employs an additional dimension into which the membrane curves. It's a good depiction, not least because it lets us visualize the way space opens up a little in the vicinity of a massive body. But as far as we know, the real world is just four-dimensional, and four-dimensional spacetime is curved all by itself, without any need for a fifth dimension.[3]

Einstein's field equations, $G_{\mu\nu} = 8\pi G_N T_{\mu\nu}/c^4$, amount to ten differential equations for the ten functions in the metric tensor. What the Einstein equations say overall is that mass, energy, momentum, pressure, and shear stress (all of which sit inside $T_{\mu\nu}$) cause spacetime to curve. In situations where all massive bodies are moving slowly and pressure and shear stress can be neglected, the most important component of the Einstein field equations is the one that relates to time only: $G_{00} = 8\pi G_N T_{00}/c^4$. We write G_{00} now instead of $G_{\mu\nu}$ because we're focusing on the Einstein equation which is labeled by $\mu = 0$ and $\nu = 0$, and conventionally we think of setting a tensor index to 0 when we refer to the time direction, whereas $\mu = 1$, 2, or 3 would refer to our usual three spatial directions. When we deal with ordinary gravity, the equation $G_{00} = 8\pi G_N T_{00}/c^4$ boils down to the rule for calculating the lapse function that we mentioned earlier. In other words (simplifying a little), the 00 Einstein equation is all you need to describe ordinary gravity. The nine other equations come into play in more extreme situations, like inside a collapsed star or near a black hole.

---

3 Many modern theoretical developments hinge on the presence of a fifth dimension, or several extra dimensions, but it is still true that intrinsic curvature of four-dimensional spacetime is what matters for the gravity of everyday experience.

In a nutshell, the two main pillars of general relativity are Einstein's field equations and the principle of optimal proper time. Using a common phrasing, matter tells spacetime how to curve through Einstein's equations, and curved spacetime in turn tells matter how to move by means of the principle of optimal proper time. Analogously, electric charges tell the electromagnetic field what to do through Maxwell's equations, and the electromagnetic field in turn produces forces on electric charges.

There's one more phenomenon that the analogy with electromagnetism reminds us to think about: radiation. Just like Maxwell's equations, Einstein's field equations have solutions that describe a self-sustaining cascade of field disturbances that propagate through spacetime. In electromagnetism, these field disturbances are electric and magnetic fields. In general relativity, they are distortions of spacetime which can be most easily described as stretching in one spatial direction while squeezing in another. Gravitational waves are produced by moving matter, much as light can be produced by moving electric charges. Once produced, they travel through spacetime at the speed of light. In effect, they are ripples in spacetime, similar to waves on water.

Like light, gravitational waves carry energy. They have been detected indirectly in compact star systems known as binary pulsars. This detection, by Russell Hulse and Joseph Taylor, was awarded the Nobel Prize in Physics in 1993. What they actually observed is a slow decrease in the orbital period of the binary star system: In effect, the stars are slowly spiraling into each other. The release of gravitational radiation drives that slow in-spiral, and the observed rate of the in-spiral matches the predictions of general relativity. The direct observation of gravitational waves by LIGO in

September 2015 is closely related to similar in-spirals and promises to be one of the great watersheds of twenty-first century physics.

We will delve into the details of gravitational radiation more deeply in Chapter 6. For now, let's note one key difference between electromagnetism and general relativity: Light waves do not interact with one another, but gravitational waves do. For example, two light waves which cross one another pass right through each other and keep on going undeflected. In contrast, two gravitational waves can collide, scatter off of one another, and head off in new directions. This tendency to scatter is extremely weak at energies accessible to us. It is so subtle an effect that it does not seem likely that anyone alive today will see a successful measurement of it. But it is undeniably part of general relativity. In fact, it is one of the main reasons we have so much difficulty merging relativity with quantum mechanics. The trouble is that the self-scattering of gravity waves becomes strong at very high energies, and we don't know how to formulate quantum theory in the presence of such strong self-scattering. String theory solves this problem in a beautiful way, but to discuss it would take us too far from our main goal. Instead, with general relativity in hand, let's learn about black holes!

CHAPTER THREE

THE SCHWARZSCHILD
BLACK HOLE

HAVING LEARNED THE BASICS OF SPECIAL AND GENERAL relativity, we are now ready to tackle black holes head-on. We will start with the Schwarzschild black hole. In the briefest of terms, the Schwarzschild black hole describes the response of spacetime to a point mass. By using the term "response," we are alluding to the idea that matter tells spacetime how to curve, and it does so through the Einstein field equations, $G_{\mu\nu} = 8\pi G_{N} T_{\mu\nu}/c^4$. A curved spacetime is mathematically described by the ten functions of the metric tensor. The Einstein field equations say that not any old set of ten functions suffice; instead, functions that vary throughout space and time in just the right way, so as to form a *solution* to the field equations, are required. The German astronomer and physicist Karl Schwarzschild published his famous solution in 1916, though in a letter to Einstein in December of 1915 it was clear he had already found the solution then, very shortly after Einstein had fully formulated his field equations.

The Schwarzschild solution is remarkably hard to understand. Even Einstein seemed not to grasp some of its essential points, in particular the smoothness of the horizon. What is this black pearl of gravity, which fell so easily into Schwarzschild's hands but which even Einstein struggled to fully comprehend?

It took approximately 50 years after Schwarzschild's discovery of his solution before a clear picture emerged of its physical significance. We have already explained some aspects of that picture—notably, the idea of an event horizon from which no signals can escape, and of an interior in which time points radially inward. For astrophysics, it is very important to understand orbits of massive objects caught in the gravitational pull of a Schwarzschild black hole, and we will devote a significant fraction of this chapter to describing these orbits and what they look like to a distant observer. We will also explain as well as we can (in the absence of any experimental verification!) what we think must happen to an object which falls into a Schwarzschild black hole. Finally, we will discuss two surprising features of Schwarzschild black holes: white holes and wormholes, which are probably not relevant to black holes formed by the gravitational collapse of old stars but are nevertheless part of the modern understanding of Schwarzschild's solution. Before we get to all of this, we will first try to directly answer the question: What is the Schwarzschild metric itself?

Far from the horizon, the Schwarzschild metric is very nearly the Minkowski metric, which we described in Chapter 1. In other words, far away spacetime is almost flat, and observers there can adequately describe their motion, as well as relative motion effects such as time dilation and length contraction, using only special relativity. Moving closer to the horizon, time slows down on account of gravitational time dilation as explained in Chapter 2. As indicated in the Preface, time changes character completely at the horizon, but since that's a complicated story, let's restrict our attention for now to the region of spacetime outside the horizon. There, the slowing down of time can be described completely by a lapse function, which is one component of the Schwarzschild metric. The rest of the Schwarzschild metric describes the three-dimensional curved space that gravity creates around the black hole. We can think of the three dimensions of space as radius plus two angular directions. Moving in the radial direction means going straight up, away from the black hole, or straight down toward it. Moving in an angular direction means going around the black hole at constant radius.

It's possible to feel confused about what radius means in the Schwarzschild solution, because you can't very well measure from the center of the black hole, which is a singularity behind the event horizon that destroys anything that touches it. The right way to think about radius is to measure the circumference of a circle centered on the singularity. This circle can be wholly outside the event horizon, on the horizon, or even inside it. Outside the horizon, we can imagine a thought experiment that would allow us to measure the circumference. It would involve a great many observers, call them Alice, Bob, Bill, Bruce, Barney, and so on through Burt. Each of them uses a rocket ship to hover at points all around the circle. We give a laser to each observer, and we also give a stopwatch to Alice. We instruct Alice to send a laser pulse to one of her neighbors (say, Bob) and at

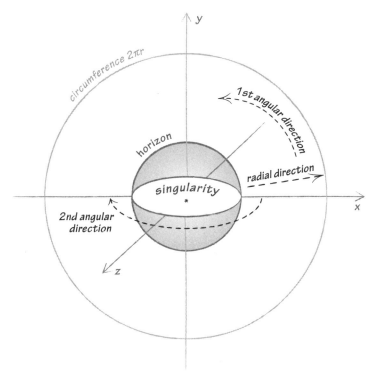

FIGURE 3.1. The radial and angular directions in the Schwarzschild solution. Outside the horizon, these three directions are the three dimensions of space. Radius is defined so that a circle centered on the black hole has circumference $2\pi r$.

the same time to start her stopwatch. As soon as Bob receives Alice's laser pulse, he immediately turns around and sends a similar laser pulse to Bill. Then Bill sends a pulse to Bruce, and so on around the circle. In the end, Burt sends a signal to Alice, and when she receives it, she stops her stopwatch. Multiplying the total time recorded by Alice's stopwatch by the speed of light gives us a length which we can reasonably

THE SCHWARZSCHILD BLACK HOLE

define as the circumference of the circle. The radius is then calculated as the circumference divided by 2π.

With our careful definition of radius as circumference divided by 2π in hand, we can revisit a phenomenon that we first described in Chapter 2: Space opens up a bit in regions where time runs slower. Suppose we have a black hole containing exactly 1 solar mass, so that the horizon is at a radius of 3 kilometers. Now let's consider two circles centered on the singularity, one at a radius of 10 kilometers and another at a radius of 10 kilometers plus 1 meter. Per the discussion of the previous paragraph, what we mean by "a radius of 10 kilometers" is that the first circle has a circumference of 2π times 10 kilometers; similarly for the second, slightly larger circle. In flat space, these two circles would be exactly 1 meter apart, meaning that to go radially outward from the first to the second, you would need to travel 1 meter. In the Schwarzschild solution, you need to travel somewhat farther than 1 meter—about 1.2 meters in total, starting from the 10 kilometer circle. Now, here's a neat fact. Time slows down due to gravitational redshift in Schwarzschild's solution by exactly the same factor that describes the stretching of the radius. In other words, the lapse function, which describes how fast time flows, is perfectly correlated with another metric function describing how much extra distance you have to travel going radially outward compared to the distance you would travel if space were flat.

In the preceding few paragraphs, we have explained almost every aspect of Schwarzschild's original solution. There's just one thing left to describe, and that is the precise form of the lapse function. Far from the horizon, it is 1, signifying that time runs at the standard rate we're used to in flat spacetime. The lapse function is 0 at the horizon,

CHAPTER THREE

meaning that ordinary time stops running there—in fact, that is one way to understand what the horizon is. In between, the lapse function varies smoothly between 0 and 1. How does it do that? The answer is that the lapse function is the square root of 1 minus a constant divided by the radius. That's a bit of a mouthful, so let's write it down: $N = \sqrt{1 - \frac{r_s}{r}}$, where N is the lapse function, r is the radius, and r_s is the radius of the horizon, called the Schwarzschild radius. Up to some factors, the Schwarzschild radius is equal to the mass of the black hole. All this detailed information is what Schwarzschild was able to extract by solving Einstein's equations.

An uncomfortable feature of the Schwarzschild solution is that the lapse function goes to 0 at the horizon, and correspondingly the radial stretching becomes infinite. This apparently singular behavior was long thought to indicate a pathology in the Schwarzschild metric. In fact, the pathology lies in the coordinates that we have chosen to describe time and radius. These coordinates are best suited to describe observers that hover at a fixed location outside the horizon. The lapse function that we discussed describes the gravitational redshift, again for observers who hover at a fixed location. That the lapse function goes to 0 on the horizon actually tells us it's impossible to hover on the horizon of a black hole. It's no wonder that the metric looks singular from an impossible perspective! If one were to describe the Schwarzschild metric from the perspective of an observer falling freely into the black hole, there would be nothing singular or unusual at the location of the horizon. The difference between a hovering observer and a freely falling one can be captured by a change of coordinates somewhat similar to a Lorentz transformation, but more sophisticated. After this change of

THE SCHWARZSCHILD BLACK HOLE

50

coordinates, which mixes time and radius, the Schwarzschild solution is seen to be perfectly smooth at the horizon. Only the singularity at the center of the black hole remains.

The Schwarzschild solution is all around us—literally! The gravitational field of the Earth is very well approximated by Schwarzschild's simple curved spacetime metric. In fact, the metric of spacetime outside any perfectly spherical distribution of mass must be given exactly by the Schwarzschild metric. Deviations from the Schwarzschild metric here on Earth (more precisely, just above the Earth's surface) arise because the Earth isn't perfectly round, it rotates, and we feel a little bit of gravitational tug from other massive bodies (notably the Moon).

If we are surrounded by a Schwarzschild metric, does this mean that a black hole horizon lurks somewhere below us, near the center of the Earth? Fortunately, no! The Schwarzschild solution only describes the spacetime geometry outside the surface of the Earth. Inside, a different solution to the Einstein field equations applies, and it is one that has no singularities (in fact, toward the very center of the Earth, the spacetime geometry becomes nearly flat). Because all planets and stars known at the time of Schwarzschild's discovery were much larger than their corresponding Schwarzschild radii, it was tempting to postulate that the properties of realistic matter would never allow a star to be concentrated into so small a volume that its radius got anywhere near its Schwarzschild radius. Though arguments were gathered over the years showing this postulate was wrong, it was not until the 1960s that the idea of black holes entered into the mainstream of theoretical physics.

A paradoxical feature of Schwarzschild's solution is that, while it purports to be the response of spacetime to a point mass, the actual point mass in question is not part of the equations that Schwarzschild's metric actually solves. To be precise, Schwarzschild's metric solves the *vacuum* Einstein field equations, $G_{\mu\nu} = 0$, which is to say there is no matter anywhere. At least, there is no matter outside the horizon. Inside the horizon, Schwarzschild's formulas can still be used, and it is still true that they solve the vacuum Einstein field equations, all the way down to zero radius. Right at zero radius, the Schwarzschild metric becomes infinite in a very nasty way. And it looks nasty from any observer's perspective, so it's a much more serious problem than the apparent singularity at the horizon we just described. We might think of this central singularity as the place where all the mass of the black hole is concentrated. But keep in mind that "place" is not even quite the right word; "time" would be better, because inside the horizon, as we'll explain more below, radius is time. Most likely, general relativity and even geometry itself fail to provide a good description of gravity very close to this central singularity, and some other theory, such as a quantum theory of gravity like string theory, is needed.

Let's review our description of black holes so far. Schwarzschild's solution to the Einstein field equations answers the question of how a point mass should curve spacetime, and the upshot is that spacetime forms a black hole. Far from the black hole, spacetime is only mildly curved, and we can understand what's going on in terms of the lapse function, recovering Newtonian gravitational physics from the way that time flows slightly faster as you move ever farther from the black hole. This approach breaks down completely at the Schwarzschild radius, where time, as measured by an observer hovering there, stops relative to a distant observer's

time. At first people thought this was a flaw in Schwarzschild's solution, or even Einstein's theory itself, but it was
eventually realized this behavior just tells us it makes no sense
to have an observer sitting exactly on the horizon. Proceeding farther inward, one eventually encounters a curvature
singularity that to this day we do not understand fully. Our
plan continuing through this chapter is to further explore the
physics of Schwarzschild black holes by asking what happens
to observers and objects who move around it or fall into it.
We will even consider the destructive region close to the
singularity.

Let's begin with a little history that actually predates
Schwarzschild's discovery. Einstein knew that an outstand-
ing puzzle in astronomy was the precession of the perihelion
of Mercury. Mercury's orbit is slightly elliptical, as permitted by Kepler's laws and Newtonian gravity. The *perihelion*
of the orbit is the point of closest approach to the Sun. But
the long axis of the ellipse, and likewise then the perihelion, slowly rotates about the Sun, or *precesses,* in the same
direction that Mercury orbits. This sort of precession had
been studied extensively, and it was understood by Einstein's
day that it could mostly be explained by the influence of
other planets. The trouble was that a very slight discrepancy remained even after accounting for all the effects in
Newtonian gravity that could be found. To emphasize how
slight this discrepancy was (and how accurate nineteenth-
century astronomical observations had become!), let's quote
the numbers. Mercury's orbit actually precesses by just over
574 arc-seconds per century, and Newtonian mechanics can
account for approximately 531 arc-seconds per century. The
remaining 43 arc-seconds per century constituted the discrepancy. The discrepancy of 43 arc-seconds per century

CHAPTER THREE

amounts to a change in the long axis of the ellipse of Mercury's orbit of approximately 1/35,000 of a degree per orbit. Before Schwarzschild found his exact solution of Einstein's field equations, Einstein was able to find a sufficiently good approximation of the Schwarzschild solution to give quite a precise account of planetary motion in the gravitational field of the Sun. When applied to Mercury's orbit, his calculations revealed a result consistent with its famous anomalous precession. Einstein had many insights, as well as some false starts, on the way to the final form of the field equations in 1915, but this result was certainly a "Eureka!" moment, giving him confidence that he had indeed discovered the correct relativistic theory of gravity.

With the exact Schwarzschild metric in hand, we can find all sorts of orbits of massive bodies around black holes that differ far more from Newtonian gravity's ellipses than the subtle precession seen in the orbit of Mercury. And yet, Einstein's early calculation contains the seed of the main idea for characterizing many of these orbits. Let's abandon the solar system altogether and head to the center of the Milky Way, where a gargantuan black hole lurks. This monster contains about 4 million solar masses. It's not really a Schwarzschild black hole; instead, it is a spinning Kerr black hole, which is a more complicated object that we'll describe in Chapter 4. For the current discussion, let's use some poetic license and pretend that the Milky Way monster actually is a Schwarzschild black hole; also we'll ignore any other matter that might be found in its vicinity. Its Schwarzschild radius is about 12 million kilometers. Our intrepid observers, Alice and Bob, decide to park their spacecraft in a circular orbit at a radius of 150 million kilometers from the black hole. That's the radius of Earth's orbit about the Sun. Because the pull of the black hole is so much stronger than the Sun's pull (about 4 million times stronger!), the circular orbit that Alice and Bob occupy is much quicker than the Earth's year-long circuit around the Sun. In fact, one full circuit takes Alice and Bob about 4 hours. At this location, gravitational time dilation makes their clocks run 4% slower than a very distant observer's.

Next, Alice boards a small shuttle craft. After undocking from the main spaceship, her plan is to briefly fire her engines to slow down her angular motion, and then turn off her engines and enjoy a wild ride. Bob, meanwhile, promises to stay put in the main spaceship and watch what happens. To help Bob track her progress, Alice mounts a flashing light on her shuttle. It flashes yellow once per second.

The point of this thought experiment is that once her engines are turned off, Alice's trajectory will be a geodesic in the Schwarzschild black hole geometry. Because her initial speed is slower than the speed it takes to maintain a circular orbit, her orbit will surely dip some way toward the black hole. In fact, if Alice fired her engines so hard that she came to a complete stop, then her subsequent motion would be to plunge radially inward and be swallowed up by the event horizon. Alice is clearly a bit of a daredevil, but she doesn't care for that scenario. So she retains some initial angular velocity and expects to whip around the black hole and then coast back out to her original radius. Once there, she can navigate back to the main ship, or she can decrease her angular velocity some more and go again.

Here's the first important point about Alice's orbits: They will precess like crazy, and the lower she causes her orbit to dip, the more her orbit precesses. Even a modestly elliptical orbit will precess a lot more than Mercury's orbit around the

CHAPTER THREE

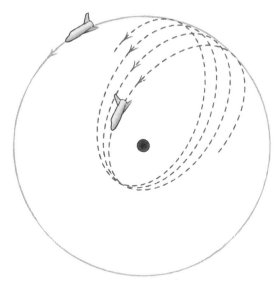

FIGURE 3.2. A circular orbit (solid curve), and a segment of a relativistic precessing elliptical orbit (dashed curve). A nonrelativistic Newtonian elliptical orbit (not drawn) would appear as a single ellipse; relativistic precession can thus be thought of as causing this otherwise static ellipse to continually rotate with time, as shown.

Sun, because in absolute terms gravity is far stronger where Alice and Bob have ventured than it is anywhere in the solar system. Despite being more pronounced, the precession effects that Alice and Bob can observe find their explanation in calculations similar to the ones that Einstein did to understand Mercury's orbit.

After some experimenting with different orbits, Alice eventually discovers that she can make her orbit precess as much as she wants. Here's how. Alice starts at the main spaceship, and she carefully adjusts her initial speed so that her orbit will dip down to a minimum radius slightly larger

THE SCHWARZSCHILD BLACK HOLE

than twice the Schwarzschild radius. Then she cuts her engines. If she arranges her initial velocity just right, she will zoom down, whirl around the black hole many, many times, and then zoom back up to the radius where she started. The scientific term for this type of motion is a "zoom-whirl orbit." Alice considers zoom-whirl orbits the ultimate rollercoaster ride. It's free-fall all the way, and in the whirling phase she is going quite fast, about two-thirds the speed of light. The trouble is, Alice is playing a dangerous game. If her initial speed is just a little too small, such that her orbit dips down below twice the Schwarzschild radius, then she will fall into the black hole unless she uses an emergency blast of her shuttle engines to accelerate back outward and escape. That emergency blast must be applied before Alice reaches the horizon, or else all is lost.

After a few runs at zoom-whirl orbits from high elevation, Alice tries to coax Bob down into a closer orbit, so that he too can feel the rush. Bob, however, is a conservative fellow, and the only orbits he is willing to try without his rocket firing are circular orbits. Alice discovers something rather odd: The lower Bob goes, the farther she has to stay outside the black hole on her zoom-whirl orbits without starting to fall in (thus requiring an emergency blast of her rockets to escape). Finally, as Bob reaches a radius equal to three times the Schwarzschild radius, Alice's game of darting in and coasting back out stops working altogether. Even if her initial speed is only slightly less than Bob's at the start of her orbit, she will wind up caught in the gravitational well and will be obliged to use her rockets to avoid getting swallowed by the horizon. Bob is now at what is called the *innermost stable circular orbit* (ISCO) of the black hole. Circular orbits of smaller radius exist, but all of them are unstable,

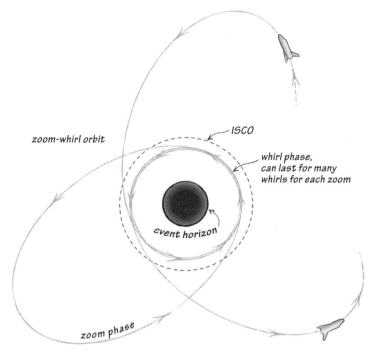

FIGURE 3.3. A zoom-whirl orbit. ISCO stands for the innermost stable circular orbit.

meaning that the slightest perturbation could send the orbits plunging into the black hole. Bob declines to fly these smaller orbits. They are essentially the orbits that Alice was flying during the whirling part of her zoom-whirl game.

Now let's ask what kind of signals Bob sees from Alice's flashing yellow signal light while she is going through all her orbital acrobatics. It helps first to take a step back and discuss an effect known as *Doppler shifting*, which happens even in the absence of gravity. In fact, the Doppler effect doesn't even require special relativity. When you

THE SCHWARZSCHILD BLACK HOLE

hear an ambulance approaching with its siren going, there's a perceptible downward shift in pitch as it passes you. To keep things simple, imagine instead of a normal siren that the ambulance emits a pure tone of definite pitch: say, A above middle C, which is 440 hertz, meaning 440 sonic vibrations per second. Suppose the ambulance is traveling at one tenth the speed of sound. (This is quite fast for an ambulance, but possible: about 77 miles per hour.) What you will hear as the ambulance approaches is a pitch approximately 10% higher than 440 hertz. Once it passes, the pitch you will hear is approximately 10% lower than 440 hertz. This change in pitch is the Doppler effect. It arises because, as the ambulance approaches you, each successive sonic vibration is produced a little bit closer to where you stand than the previous one. Therefore those sonic vibrations are a little bit "crowded" together as they move toward you, and they get to you in quicker succession than if the ambulance were standing still. A similar phenomenon occurs with light in special relativity. If, in the absence of any gravitational field, Alice flies directly toward Bob, then the electromagnetic vibrations that comprise the yellow light from her signal will appear to him more crowded together—shorter in wavelength, and correspondingly higher in frequency. That means that the yellow light will appear somewhat blueshifted. If instead Alice flies away from Bob, her yellow signal light will appear to him somewhat redshifted. By the same token, flashes that Alice sends every second (according to her clock) will be seen by Bob more frequently than once a second if Alice is moving toward him, and less frequently if she is moving away from him. It seems like the Doppler effect should get mixed in with the effects of time dilation in special relativity, and that does

CHAPTER THREE

indeed happen; but the upshot of a fully correct relativistic treatment is what we have described.

When Alice is performing zoom-whirl orbits, gravitational time dilation has a redshifting effect on light traveling up to Bob from her shuttle which is over and above the Doppler shift. In contrast, if Bob were to send a light ray down to Alice, it would get gravitationally blueshifted. These gravitational effects are entirely due to the variable rate of time flow at different depths in a gravitational well, and they are variants of the phenomenon behind the Pound-Rebka experiment explained in Chapter 2. An additional complication is that photons from Alice's signal light can travel in more or less complicated trajectories on their way to Bob. The most straightforward scenario is for Alice to be directly below Bob when her signal light flashes, just as she is in the middle of a whirl around the black hole. Then photons going more or less straight up will reach him, somewhat redshifted due to gravitational effects.[1] However, if Alice is on the opposite side of the black hole from where Bob is when her signal light flashes, photons can still get to him, but they have to dodge around the black hole first! Amazingly, this can happen. Einstein even anticipated it. Indeed, another early confirmation of relativity—the one that elevated Einstein to worldwide fame—was the observation of the deflection of starlight by the Sun during a solar eclipse in 1919. This deflection is a weaker version of the way photons from Alice's shuttle can sneak from her location behind the black hole half way around the horizon to

1 Special relativistic time dilation also plays a role in redshifting photons of the type that go straight upward from an orbit at nearly constant radius. Ordinarily, our intuition is that time dilation is folded into the Doppler effect, but it's not quite that way for photons emitted at right angles to the motion of the source.

THE SCHWARZSCHILD BLACK HOLE

reach Bob. But that's not all that they can do! Like Alice herself, photons from her signal light can find special orbits that take them all the way around the horizon, even several times around the horizon, before finding their way upward to Bob. In principle, photons can even circulate endlessly around the black hole at a radius equal to one and a half times the Schwarzschild radius.[2] This circular orbit, called the *light ring*, is unstable; nevertheless, this property of a black hole is responsible for the bright ringlike feature outlining the black hole's "shadow" that radio telescopes are currently searching for, as we'll briefly review at the end of Chapter 5.

In sum, Bob will see each of Alice's light pulses with some combination of gravitational redshifting and Doppler red- or blueshifting; moreover, he will see weak echoes of each pulse, corresponding to light that traveled around the black hole once, or even more times, before escaping outward. The maximum redshifting will be sufficient to push the photons altogether out of the visible spectrum, well into the infrared, while the maximum blueshifting will be just enough to put the photons far on the blue side of the visible spectrum. In short, Bob will see all the colors of the rainbow!

We've skirted the horizon of a black hole long enough. It's time to cross it. Alice and Bob each volunteer the other to cross, but, wisely, they both decline, and instead they send in a satellite probe. To keep things simple, they go back to the 150 million kilometer radius of their original

2 A photon's behavior is somewhat different from the behavior of Alice and her shuttle because photons are perfectly massless. In particular, a photon's circular orbit at one and a half times the Schwarzschild radius is the only circular orbit, in contrast to the many possible circular orbits, both stable and unstable, available to Alice and her shuttle.

CHAPTER THREE

circular orbit, and they hover there while they launch the probe in such a way that it starts from rest and plunges radially downward into the black hole without any zoom-whirliness. They attach Alice's flashing yellow signal light to the probe so that they can watch what happens. Because of a combination of gravitational and Doppler effects, the pulses arrive less frequently than once a second, and the light is redshifted. The free-falling probe will measure a time of 2,638 seconds to reach the ISCO from its initial position of 150 million kilometers, and then an additional 122 seconds to reach the event horizon. At this moment, at least according to classical general relativity, it will cross the horizon without any fanfare. In fact, there's nothing special there to let it know it has crossed. Alice and Bob, however, will *never* see it cross, since the gravitational time dilation becomes infinite approaching the horizon. Said another way, each subsequent pulse from the navigation light takes longer and longer to reach them, and at some time, a last pulse will come from just outside the horizon. This is true no matter how short they set the interval between pulses—but let's stick with the plan of one pulse per second as measured by a stopwatch on the probe. Suppose we arrange things so that one of the probe's light pulses is emitted exactly as it crosses the horizon. This pulse will never be seen by Alice and Bob, but all the previous ones will be. The last pulse they see arrives 3,741 seconds after the probe left, while the second-to-last arrives 3,686 seconds after. Thus, from Alice and Bob's perspective, the one-second pulse interval has stretched to 55 seconds between the second-to-last and last pulses. Their wavelengths would furthermore be redshifted by factors of 79 and 40 for the last and second-to-last pulses. If these pulses were emitted by the probe as bright yellow light at

THE SCHWARZSCHILD BLACK HOLE

a wavelength of 570 nanometers, they'd be measurable by Bob and Alice as infrared photons with wavelengths of 45 and 23 microns, respectively.

The gravitational time dilation (which we've also referred to as gravitational redshift) is proportional to the inverse of the lapse function, which goes to 0 at the horizon; hence the time dilation goes to infinity there. That's one way of understanding why the pulse the probe emits at the horizon never makes it to Alice and Bob, let alone any of the pulses it emits afterward: Inside the horizon, the time dilation goes "beyond infinity." But what does a statement like that even mean? The free-falling probe doesn't notice anything strange at the horizon. But if the probe tries to accelerate back out from behind the horizon, it will fail. No matter how strong a boost it uses to try to get back out, it cannot even reach the horizon. It can't even stop its inward motion. This is the fateful feature of the black hole interior that we emphasized in the Preface. Moving forward in time means moving inward in radius. No force can pull an object back out of a black hole, any more than a force can pull us backward in time. Neither can the photons from the probe do anything other than fall inward, once the probe has crossed the horizon. Time dilation has indeed gone "beyond infinity" in the sense that time inside the black hole is altogether other than time outside. Time inside the horizon runs inward, and the future there is limited by the singularity.

Inward-pointing time inside the horizon is so central to black hole physics that we will circle back to the language of differential geometry to give a more precise account of it. Recall that a spacetime metric has two tasks: It tells us the proper time between events that are timelike separated, and it tells us the proper distance between events that are spacelike separated. There is a beautiful way to write a single formula for the spacetime metric which accomplishes both tasks in one go. The trick is to write a formula not for distance, but for the square of the distance between two nearby events. If the square of the distance is positive, then those two events are spacelike separated. If instead it is negative, then they are timelike separated, and what we thought was the square of the distance is really minus the square of the proper time between the events. In the Schwarzschild solution, as in any solution to Einstein's equations, the formula for the metric (based on the lapse function, radial stretching, and so forth) is really one of these distance-squared formulas which can return either positive or negative values. Two events separated slightly in the radial direction have a distance squared between them which is positive outside the horizon, but *negative inside the horizon*. This last point is the key: Negative distance squared means that events are timelike separated. In other words, radius becomes timelike, while time becomes spacelike. Strange as all this sounds, there's nothing funny in the curvature of the Schwarzschild geometry going on here; instead, one's usual notions of time and radius get rearranged upon crossing the horizon.

Notwithstanding the way time and radius get mixed up inside the horizon, our original idea of radius in the Schwarzschild solution is still valid: Even inside the black hole, radius can still be understood as the circumference of a circle centered on the origin, divided by 2π. To put it another way, the area of a sphere at any given radius in the Schwarzschild solution is 4π times the square of the radius— the same formula we learn in school. The true meaning of this formula inside the black hole is alarming: The radius there, we have now learned, is also time, so the sphere we're

THE SCHWARZSCHILD BLACK HOLE

talking about is the total extent of space in the two angular directions at a fixed time. As time moves forward (meaning that radius moves inward), the sphere gets smaller. And smaller and smaller. And then, "Ouch!" The singularity.

To explain the sort of "ouch" we would feel as we head toward the singularity, we need to explain tidal forces. As Newton knew well, the ocean tides that we see here on Earth are a manifestation of the Moon's gravitational tug on the planet.[3] The Moon does this by pulling a little more strongly on the near side of the Earth than on the far side. This uneven pull causes the Earth to elongate just slightly in the direction of the Moon. The whole Earth feels this elongating pull, but the oceans respond to it more, because water has no stiffness. The tidal forces from the Moon act as if they were pulling the far side of the Earth *away* from the Moon while at the same time pulling the near side of the Earth toward the Moon. This is at first very counterintuitive, because we know that gravity is purely attractive. The point is that tidal forces are the residual effect after we account for the average pull of the Moon on the Earth. This average pull slightly alters the orbital motion of the Earth, whereas the tidal forces stretch the Earth just a little.

As the probe falls through the horizon (see figures 3.4 and 3.5), it is already in principle experiencing some tidal forces, but they are not particularly strong because the black hole is so enormous, and the probe itself may be rather small. Let's say it's just a meter across. Things change quickly inside the black hole. As already discussed, no amount of acceleration will allow the probe to avoid the singularity once it's inside

3 The gravitational pull of the Sun also influences tides, but for simplicity we ignore it here and focus on the stronger effects from the Moon.

CHAPTER THREE

the horizon. In fact, it turns out that if we want the probe to maximize its proper time before its doom, the best we can do is to have it not accelerate at all, but stay on a geodesic. It will then hit the singularity about 27 seconds after crossing the horizon. Tidal forces from the black hole's gravitational pull will increase rapidly as the probe approaches the singularity, and by about 10 to 100 microseconds before hitting the singularity (the exact amount depending on how strong a metal we use to construct the probe) its hull will fracture. The increasingly powerful tidal forces pull apart the pieces of the probe into ever smaller, thinner bits, and then even these thin bits are stripped into their constituent atoms. But it doesn't stop there. Soon the tidal force grows strong enough to strip all the electrons off the atomic nuclei, then rip the nuclei apart into free protons and neutrons, then into quarks and gluons. Ouch indeed! We cannot say what comes next, because as far as we know, quarks, gluons, and electrons are pointlike. But what we can say is that the two angular directions of three-dimensional space are themselves getting squeezed ever smaller as we approach the singularity, while the third spatial direction—corresponding to what we used to call time outside the black hole—undergoes ever more radical stretching. Correspondingly, everything, including our probe, gets squished and stretched into an infinitesimally thin line.

It seems that we have now explored the Schwarzschild solution from beginning to bitter end. Truly it is a marvel, characterizing in a simple and precise way the curved spacetime geometry we live in, and at the same time giving an approximate account of the spacetime geometry surrounding the heaviest object in our galaxy, namely the gargantuan black hole at its core. In and of itself, a Schwarzschild black

THE SCHWARZSCHILD BLACK HOLE

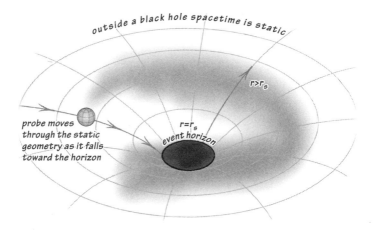

FIGURE 3.4. Probe falling into a black hole: view from outside the event horizon.

hole is perfectly static, lurking like a spider at the center of a curved geometrical web. As we've learned, objects which fly too close to the horizon must struggle mightily to escape, and anything that crosses the horizon is (we think!) soon tidally digested into the thinnest imaginable stream of matter arcing into the singularity.

In fact, this is not the end of the story of the Schwarzschild solution. There's another face to the Schwarzschild metric that's the diametric opposite of the black hole part of the spacetime. It's called a *white hole*. Here, beginning at a singularity, the flow of time drags all of space away from the singularity, expelling everything across a one-way boundary to the outside. Once kicked out, you can never get back into the white hole. The reason the white hole must be part of the Schwarzschild solution can be gleaned from the following apparent paradox. Geodesics, barring singularities,

CHAPTER THREE

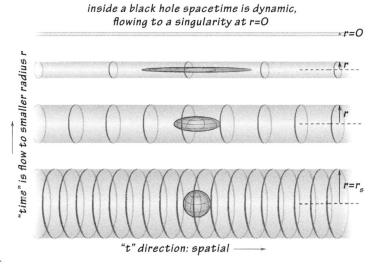

inside a black hole spacetime is dynamic,
flowing to a singularity at r=0

"time" is flow to smaller radius r

"t" direction: spatial

FIGURE 3.5. Probe falling into a black hole: view from inside the event horizon. Inside, the probe is swept along in the collapse of spacetime. As "time" advances from $r = r_s$ at the event horizon toward $r = 0$ at the singularity, the probe stretches in one spatial direction ("t") to ∞ and compresses in the two spatial spherical directions to 0.

are what is technically called "complete:" As trajectories of optimal distance through spacetime, they never just begin or end—for a particle or photon geodesic, there is always a path forward or backward in time from any point. The only place where this property may break down is if a geodesic runs into a singularity; then a theory of quantum gravity would be needed to understand what happens. Of course, nongravitational forces can cause a particle to move along a trajectory that's not a geodesic, but as a path through spacetime the geodesic is always there. For example, if you are sitting on a chair in your favorite coffee shop reading this book, you are not following a geodesic: The pressure that

THE SCHWARZSCHILD BLACK HOLE

the chair and ground exert on you prevent this. But there is still a geodesic, moving through the ground toward the center of the Earth, that some particle or object not subject to this pressure force, such as a neutrino, would follow.

With all that as background, here's the apparent paradox in the Schwarzschild spacetime. A photon that Alice and Bob see that traveled radially outward from the probe which they launched into the black hole follows a geodesic after its emission by the light source on the probe. But the geodesic it follows is complete, so it extends past this point of emission, all the way back to the Schwarzschild radius, and beyond. Think of it this way. Suppose we followed the path of the photon backward in time from the point where Alice and Bob see it. It's moving to smaller radii, and at some time back in the past it reaches the location of the probe. Here the photon's life ends, as this is where the probe created it. But the geodesic the photon was moving along doesn't end: It continues to smaller radii going farther back in time, and nothing in principle prevents an actual photon from having followed this path. The path continues all the way to the horizon, and since the Schwarzschild radius is not a singularity, it even crosses it and extends to smaller radii than that. We know that nothing can escape the black hole. But it seems this geodesic offers a path of escape, so we have a contradiction.

The resolution to this apparent paradox is that this geodesic does not come from the black hole part of the Schwarzschild metric, but from an entirely different part of the spacetime—the white hole—where the dynamical flow of spacetime is exactly opposite to the black hole. Inside the white hole, where (or really "while") the radial coordinate is less than the Schwarzschild radius, going forward in time

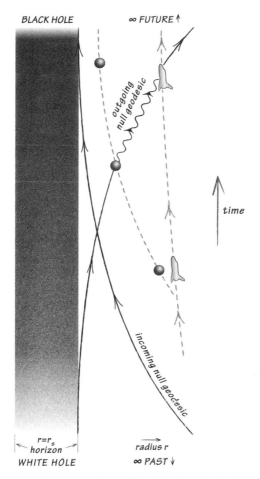

FIGURE 3.6. A depiction of the full black hole/white hole nature of the Schwarzschild solution. From an outside observer's perspective (in this case, that of the shuttle observers), all radially outgoing null geodesics originate from the white hole in the past; all radially incoming null geodesics fall into the black hole in the future. However, neither class of geodesic is ever "seen" to cross the white hole horizon or the event horizon, as that only happens in the infinite past or future as judged by the observer. An outgoing photon emitted by the probe jumps onto an outgoing null geodesic track until it is absorbed (i.e., observed) by the shuttle.

THE SCHWARZSCHILD BLACK HOLE

means moving to larger radii. So instead of being trapped in this region, anything inside is forced out with no chance of getting back in. Zero radius inside the white hole is a singularity, but the tidal forces are opposite what they were for the black hole: Thin lines are squished and stretched back into spheres.

Where is this white hole? More precisely, when is it? The answer is that the white hole happened in the past; in fact, it was happening arbitrarily far in the past. In a similar sense, the black hole is part of the future, and it will persist forever (ignoring quantum effects). If this is confusing, another way to think about this white hole/black hole business is to draw an analogy with Big Bang cosmology. In Einstein gravity, the Big Bang is a singularity where our universe "began." (As with the singularities in black holes, general relativity breaks down at the Big Bang singularity, so what really happened there is a mystery.) Even though we can see the photons that came from the Big Bang all around us, forming the cosmic microwave background, of course we can never go to the Big Bang—it's not a place, rather it was a time in our past, and what we're left with is an expanding universe. Likewise, taking some poetic license, we can think of Schwarzschild's solution as describing a spacetime that began its life as a white hole, and what's left is a black hole. From outside, any photons that were produced in the white hole will look like they're coming from the region of space where the black hole is now. And they did, but when they were streaming outward toward the Schwarzschild radius to eventually cross it, the black hole wasn't there yet, only the white hole.

To hone our understanding of the Schwarzschild black hole, we've explored many possible paths for massive objects

CHAPTER THREE

(like Alice's shuttle or the ill-fated probe) and for photons. Together, all such paths are termed "causal paths," because it's possible for one spacetime event to influence a second event if both lie on a causal path with the first preceding the second. If we enlarge our view of spacetime a little by considering paths that are not causal, then one finds another remarkable feature of the Schwarzschild solution: a so-called wormhole, or Einstein-Rosen bridge, that connects the outside world that Alice and Bob live in to a second outside world with an identical geometric structure. In that world, like-minded adventurers Alicia and Bradley could perform the same experiments on their black hole, and they would come to all the same conclusions as Alice and Bob. However, neither pair would ever know about the other's existence, because the only connections between their worlds are noncausal paths through either the black hole or the white hole interior, which no particle, massive or massless, can follow. In other words, the two outsides are causally disconnected from each other, but the interiors overlap. It is thought that a wormhole could connect extremely distant parts of the universe, and this notion is at the root of large swaths of science fiction. The trouble is that connecting distant regions of spacetime noncausally doesn't really connect them at all in any practical sense. Noncausal connection precisely means that nothing (literally, no thing) can get from one side to the other. The wormhole is nontraversable. In general relativity there are solutions to the field equations that describe traversable wormholes, but these all require "exotic" forms of matter which are not yet discovered, or which may not even exist at all. We will return to a discussion of what "conventional" wormholes, which is to say nontraversable ones, might mean at the end of Chapter 7.

THE SCHWARZSCHILD BLACK HOLE

Despite all the turmoil and violence going on inside the horizon (not to mention alternative worlds), not a whimper of it is noticeable from the outside. In fact, this is an aspect of a more general property of black holes referred to as the "no-hair theorem." In this chapter, we are focusing mainly on one particular kind of black hole, the nonrotating Schwarzschild black hole. In the next, we'll look at rotating black holes (called Kerr black holes), and black holes that have electric charge. You might be wondering then, how many kinds of black holes are there? The claim is that once you know the mass, charge, and spin of a black hole, you know the whole geometry exactly. It's this impressive claim that is called the no-hair theorem, or sometimes the *uniqueness theorem*. Here, unique means if we pick a specific value for each of the mass, spin and charge parameters, then there is one and only one shape that the black hole horizon can have. The origin of the term "no-hair theorem" comes from a humorous way of thinking about what nonunique horizon shapes might look like if they were possible. Perhaps black holes can have bumps, mountains, dimples, or valleys. What should the general term be for such hypothetical features of black hole horizons? "Hair" is the favored choice. It must be admitted that a term like "no-feature theorem" has nothing like the panache of the bold statement, "Black holes have no hair."

The way we understand the no-hair theorem intuitively is that horizons might have some temporary features, but they settle down to nothing in about the amount of time it takes light to travel once around the light ring. Proving this with true mathematical rigor is difficult. The original no-hair theorem, due to the Canadian physicist Werner Israel, is less ambitious, but it is actually a rigorously proven result. He proved that if we assume the black hole is in a steady state (meaning there are no temporary features that would need to settle down), then on and outside the horizon, a nonrotating black hole must be Schwarzschild. In other words, Schwarzschild is the unique answer to the question of what nonrotating, steady state spacetime geometries solve Einstein's equations. This result was later extended by other scientists to include rotating black holes, described by the Kerr solution that we will discuss in Chapter 4. Proving that steady state solutions are indeed unique (which is what Israel did) doesn't fully demonstrate the grander claim that all black holes settle down to Schwarzschild or Kerr solutions, but it is a step toward it.

All evidence is that the Schwarzschild and Kerr solutions are indeed the stable endpoints of gravitational collapse. When a black hole first forms from the collapse of a massive star, or when two black holes smash together, the spacetime about the horizon is definitely not in a steady state, and it has a lot of interesting structure. But very quickly all the structure gets carried away as gravitational waves, and the geometry outside the event horizon becomes this perfect, smooth, stationary shape described by an exact solution of the Einstein field equations. What happens inside the horizon is far less certain. In fact, despite our in-principle knowledge of the interior of the Schwarzschild and Kerr solutions, what goes on inside the horizon of a dynamically formed black hole is largely a mystery that scientists and mathematicians are still trying to unravel.

A black hole formed when a massive star collapses will not have a white hole in its past—the star is there instead. Nor will there be any wormhole to a second universe. Actually, there's still a bit of a mystery as to how the supermassive black holes seen in the centers of galaxies got there. It's not inconceivable that they might have something akin to a white hole in their past, or wormholes connecting them to other pieces of the universe. White hole regions in the far past of supermassive black holes in our universe would probably have to be a substantial modification of the white holes of the Schwarzschild metric, because our observable past (the Big Bang) looks quite different from a Schwarzschild white hole. It is also entirely plausible that supermassive black holes formed via collapse of massive stars very early on in the evolution of the universe and then grew by consuming matter and other black holes over time to get as large as they are today. In that case there wouldn't be any white hole or wormhole pieces connected to them. The bottom line is that there's a lot of observational evidence that black hole regions exist in our universe, but none for white holes or wormholes.

We've begun to convey some of the wonder of what a black hole in general relativity truly is. And hopefully now you can also appreciate why it took scientists so long to come to grips with what the Schwarzschild metric actually represents, even though it was available in exact mathematical detail to fully explore shortly after Einstein published his field equations. A lot of new mathematics, including the Kerr solution discovered in 1963, was needed before the Schwarzschild solution could be taken seriously. It was also crucial that astronomers began to discover objects in the universe that defied conventional explanations yet could be understood if they were black holes. Otherwise this realm of general relativity would simply have been deemed a mathematical oddity and of no physical relevance (as it had been in the early years of the theory). Most of our modern understanding of black holes crystallized after Einstein's passing, so unfortunately he could not fully appreciate how truly mind-blowing the consequences of his new theory were.

CHAPTER THREE

CHAPTER FOUR

●

SPINNING BLACK HOLES

In Chapter 3, we described all the glorious consequences of Schwarzschild's solution to the Einstein field
equations, which represents a single, static, nonrotating
black hole. Here we'll discuss an extension of the Schwarzschild solution that describes rotating black holes. This extension is called the Kerr black hole in honor of Roy Kerr,
the mathematician who discovered the solution. The Kerr
black hole is important because black holes in the universe
almost always have some rotation, or *spin*, and this produces
interesting new effects. One of the main effects of its spin
is that spacetime is dragged around the black hole as it rotates, something called *frame dragging*. This causes geodesics to exhibit a new kind of precession. Recall that in the
case of a Schwarzschild black hole, precession is the rotation of the ellipse of the orbit, but this rotation happens
within the fixed, two-dimensional plane of the orbit. In the
Kerr solution, the new twist that frame dragging adds is

that now the orbital plane itself rotates about the spin axis of the black hole, in the same sense (clockwise or counter-clockwise relative to the axis) as the rotation of the black hole. The closer to the black hole the particle is, the faster this frame-dragging-induced rotation becomes. In fact, in a zone called the *ergoregion*, the frame dragging becomes so extreme that all particles—geodesic or not, massive or not (so even photons)—are forced to circulate about the black hole in the same sense as its spin. The existence of an ergoregion also allows rotational energy to be extracted from the black hole; we'll describe one method for doing so called the *Penrose process*.

We will also briefly introduce electrically charged black holes, which are solutions of Maxwell's equations of electromagnetism as well as Einstein's field equations. Charged black holes are not so important for astrophysics, because (we think) most black holes in the universe are nearly electrically neutral. However, they illustrate some interesting ideas, in particular the notion that if a black hole contains too much charge, the event horizon ceases to exist! It is believed, however, that no physical process can cram enough charge into a black hole to eradicate the horizon; so a more correct statement is that there is a maximum electric charge that a black hole can carry. Likewise, the spin of a Kerr black hole cannot be arbitrarily large. Black holes that have the maximum possible charge or spin are called *extremal*. Though charge and spin don't change the broad properties of the spacetime outside the event horizon by much, inside is another story. There, after some time, the collapse of spacetime (which for a Schwarzschild black hole progresses all the way to a singularity) slows down and reverses at what's called an *inner horizon*. Though not a singularity, the inner

CHAPTER FOUR

horizon has some bizarre properties, one being that the field equations in a sense "fail" there and cannot uniquely predict what happens to spacetime beyond it. If one assumes that the solution can be extended as smoothly as possible beyond this inner horizon, then spacetime expands to a new region with even more bizarre properties: a negative mass singularity and trajectories along which observers can move backward in time. We explore all of these properties in more detail in this chapter.

Let's start by motivating the search for a spinning black hole. Throughout this chapter, we mean *spin* in the classical sense (i.e., rotation about a specified axis) and not in the quantum mechanical sense. Angular momentum is a measure of how much a body rotates, or spins. Both quantum mechanical and classical spin are measures of angular momentum, though they have rather different mathematical and physical characteristics. Angular momentum is an important property in physics, one reason being that for an isolated system it is a *conserved* quantity. An outside force (in the form of a torque) can change the angular momentum of the system, but as a consequence of Newton's third law, which holds in quantum mechanics and relativity, this is balanced by an equal and opposite change in the angular momentum of the agent exerting the outside force. Almost any planet, star, or black hole in the universe has at least some angular momentum. This is because of all the intricate dynamics and interactions with surrounding matter in the universe that unfold as the body forms and evolves. There is nothing new about what we're saying here—it is all classical mechanics dating back to Newton's time and earlier. But it does imply that there are certain features of generic black holes we expect to encounter in the universe that Schwarzschild's solution, describing a black hole that has exactly zero angular momentum, misses.

What we need then is a solution to the field equations describing a rotating black hole. We want to be able to recover the Schwarzschild solution as the special case when the rotation becomes vanishingly small. Given that the Schwarzschild solution was published less than a year after general relativity, it might seem strange that it was only in 1963 that Roy Kerr discovered the long-sought rotating solution. Schwarzschild assumed spherical symmetry to derive his solution, but it turns out that when the black hole spins, it distorts nearby spacetime in such a way that the geometry can no longer be spherically symmetric. Kerr looked for a less restrictive class of solutions that are called *axisymmetric*. These solutions have a single axis of symmetry about which you can rotate the geometry and not produce any change. For example, an American football is axisymmetric (ignoring the seams, the texture of the surface, and any distinguishing logos painted on it). The symmetry axis runs from tip to tip along its lengthwise direction. A well-thrown football will spin around this axis, and you won't notice the spin (other than the blurring of the logo as it whips around). Thrown with less skill, a football spinning around some other axis will look like it's wobbling or tumbling as it flies through the air. Disks and cylinders are other examples of axisymmetric geometries. A sphere is technically also axisymmetric, but it has an extra symmetry in that it looks axisymmetric about any axis passing through its center.

It turns out that if the geometry of spacetime is spherically symmetric, the field equations are drastically simpler than the less-restrictive axisymmetric situation, which is one reason it took so long for the Kerr solution to be discovered.

CHAPTER FOUR

Removing the restriction of axisymmetry complicates the field equations even more, and it's natural to wonder whether even more intricate black hole solutions await discovery. Not so, thanks to the remarkable no-hair property of black holes discussed in Chapter 3. Recall that this theorem states that any temporary features (a.k.a. "hair") that a black hole might have are lost very quickly, and the black hole settles down to a unique, stationary state. In the absence of matter or electric charges, this stationary state is the Kerr metric. In other words, any nonaxisymmetric features that a black hole might have are necessarily temporary. Stationary black hole solutions to the Einstein field equations more complicated than Kerr do not exist.

Many features of black holes aren't qualitatively affected by spin: for instance, that time dilation occurs between local and distant observers and becomes infinite approaching the horizon, that the horizon is a one-way boundary and space-time starts to collapse in on itself upon crossing it, and that orbits passing sufficiently close to the black hole can exhibit zoom-whirl dynamics. However, the *details* of these effects can be quite different, and some new phenomena arise stemming from two important ways in which spin changes the geometry outside the black hole. First, as just mentioned, the geometry is no longer spherical. In the Schwarzschild metric, surfaces of constant lapse function (meaning constant gravitational redshift) are geometric spheres. In the Kerr metric, the analogous surfaces flatten around the poles where the axis of rotation passes through and conversely bulge out along the equator. This is similar to how the shapes of the Earth, the Sun, or other massive astronomical bodies that would otherwise be spherical distort due to their rotation. The closer to the black hole's event horizon, the more noticeable this flattening/bulging is, and it also becomes more pronounced the faster the black hole spins.

The second significant way in which rotation changes the geometry is that it causes spacetime itself to start flowing around the black hole, ever faster as one approaches the horizon. We will explain in more detail what we mean by spacetime "flowing" by describing how this affects geodesic trajectories, but an apt analogy would be to think of how air flows around a tornado. Here air represents the spacetime, and the geodesics are any dust particles (or unfortunate cows) swept up by the tornado and carried around the funnel. In the spacetime context, this effect is called frame dragging. This property is not unique to black holes; in fact, the Earth's spin also causes frame dragging, but at exceedingly small levels compared to a black hole (so small it can be ignored by the GPS satellites and has only recently been measured by the sensitive Gravity Probe B and LAGEOS satellite experiments).

To start exploring all the consequences of frame dragging, let's again drop our trusty satellite probes, releasing them from rest at large distances away from a Kerr black hole, to trace its geodesic structure. For a Schwarzschild black hole, because it is spherically symmetric and nonrotating, there's nothing special about identifying a plane intersecting it through its center as the equatorial plane, and the two points on the horizon directly north and south of the equatorial plane as its poles: Any orientation of this plane is as good as any other. With a rotating object like a Kerr black hole, it's most natural to define the north and south poles as the points on the horizon connected by the spin axis of the black hole, and the equatorial plane as lying at a 90° angle relative to this axis. Because of the frame dragging and

CHAPTER FOUR

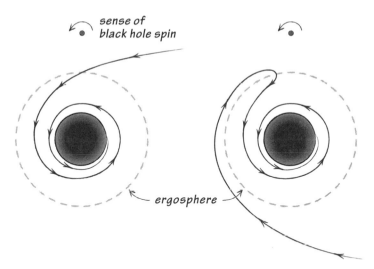

FIGURE 4.1. Effect of frame dragging on geodesics falling into a Kerr black hole. The path of the geodesic falling into the Kerr black hole is shown; on the left (right), the particle has positive (negative) angular momentum relative to the spin of the black hole.

axisymmetric nature of the Kerr black hole, it now matters at which angle relative to the spin axis the position lies from which we initially drop our probes. Let's look at the two extreme cases: one probe dropped in directly along one of the black hole's poles (whether north or south is unimportant), the other along the equator. In a Schwarzschild spacetime, there'd be no difference: Both probes would fall in radially as described in Chapter 3. The same happens in the Kerr spacetime for the probe dropped in along a pole, though the exact progression of gravitational time dilation and redshift, as seen by a distant observer as the probe falls, will differ. Along the equator it is a very different story. At first, the probe will fall in radially, but as it gets closer to the horizon,

SPINNING BLACK HOLES

the spin of the black hole will start to drag it around the black hole. As seen from a distance, its trajectory will look like a shrinking spiral in the equatorial plane, wrapping closer around the horizon in ever tighter coils but never crossing it. Light from the probe will be redshifted and time dilated as for the Schwarzschild solution, but now it would seem to be coming from a point on the horizon that's rotating with a fixed angular speed. This angular speed would be the same for probes dropped in from any angle, though they'd end up at correspondingly different positions of latitude spread over the horizon. Observing the angular motion of infalling probes is one way to measure how fast the black hole rotates.

From the probe's perspective, falling in from outside, it would also see itself start to be dragged around the black hole. As in Schwarzschild spacetime, it will reach and cross the horizon in finite time according to its watch. So there is still an infinite disparity between the rate of time flow of an observer crossing the horizon versus one far away. Moreover, by the time the probe crosses the horizon, it would have observed itself circling the black hole only a finite number of times. An outside observer never sees it cross. Instead, the probe appears to hug the horizon and rotate around forever with constant angular speed. So there's also an infinite disparity between the local and distant measurements of the amount of the probe's twisting about the spin axis of the black hole.

The effect of frame dragging on an orbit more complicated than the equatorial or polar free fall just described can be understood through the notion of orbital plane precession. About a Schwarzschild black hole, any geodesic that orbits the black hole moves in a fixed two-dimensional plane that intersects the coordinate center of the black hole.

CHAPTER FOUR

This we refer to as the orbital plane. In this plane the ellipse of the orbit precesses as discussed in Chapter 3, but it never leaves the plane. Near a Kerr black hole, frame dragging causes the entire orbital plane to rotate, or precess, about the spin axis. How fast this precession occurs depends on the spin of the black hole, how inclined the orbit is relative to the equatorial plane, and how close to the black hole the probe is. Equatorial orbits always stay in the equatorial plane, while orbits that circle about the poles of the black hole show the most orbital plane precession. An orbit that is far from the black hole will show very little precession, regardless of the inclination of the orbit or spin of the black hole. So again, as with Schwarzschild black holes, very far from the black hole the orbital dynamics is well described by Newtonian physics. At the other end of the spectrum, the orbital plane precession can be particularly pronounced for zoom-whirl orbits, especially during a finely tuned whirl. Here, instead of orbiting in a circle, the probe will trace out a pattern over a portion of a sphere between a fixed latitude above and below the equatorial plane. For an equatorial orbit this is just a circle, but a polar orbit will fill out the sphere as it whirls.

Frame dragging also influences nongeodesics (probes with rockets). Close to the horizon in the ergoregion, the frame dragging is so strong that all timelike and null trajectories are forced to rotate around the black hole in the direction of its spin. If you're outside the horizon but inside the ergoregion, even if you're blasting full force to move counter to the direction of the spin, and regardless of how large this force is, you'll still be dragged around the black hole in the direction of its spin. The shape of the boundary of the ergoregion, called the ergosphere, is a flattened version of the horizon, touching the horizon of the black hole at the poles and extending farthest outside along the equator (see figure 4.3 on page 95). The faster the black hole spins, the larger the equatorial bulging of the ergosphere becomes. Interestingly, there's a limit to how rapidly a black hole can rotate, and if spinning at this limit the black hole is said to be extremal. For such an extremal Kerr black hole, the equatorial extent of the ergosphere grows to twice the radius of the horizon. Inside the ergosphere, all particles must move in the same direction around the black hole, though it's possible for this motion to be faster or slower depending on where a particle is in the ergosphere and whether it is acted on by forces other than gravity. If we do approach the horizon, then, as seen by an outside observer, the time dilation and frame dragging work together to cause all particle trajectories, geodesic or not, to rotate around with the same angular speed as that of the horizon.

Why can't a black hole spin arbitrarily fast? Mathematically, the Kerr solutions can have more spin than extremal, but then the horizon vanishes (i.e., they are no longer black holes). These solutions are problematic in several ways, one being that without a horizon the singularity in the spacetime is revealed to the outside universe. What's wrong with that? In theory nothing, but classical general relativity cannot predict what happens to the causal future of a singularity, and so we don't even know what "revealed" would mean here. Using model calculations and simulations, scientists have tried to overspin black holes (or form singularities without horizons to begin with), but in setups mimicking what might exist out there in the universe, none have succeeded. This failure was anticipated decades ago by British physicist and mathematician Sir Roger Penrose, who formulated the Cosmic Censorship Conjecture, stating that all singularities that can form in nature are "clothed" by event horizons. From a physicist's perspective, this is an unfortunate rule if nature were to enforce it. The reason is we believe the singularities in spacetime predicted by general relativity are places where the theory breaks down, and what actually happens must be described by a new theory— call it quantum gravity. Seeing such events could give unprecedented insight into what quantum gravity truly is, but if they only happen hidden behind event horizons, we're out of luck. We will return to this subject again when discussing black hole collisions (Chapter 6).

A brief recap is in order. Rotation complicates the geometric structure of Kerr black holes compared to the Schwarzschild case, and it adds a new wrinkle to particle trajectories that are close to the horizon: frame dragging. Suppose we throw a bunch of probes with flashing lights into a Schwarzschild black hole from all directions. An outside observer will never see them cross the horizon; instead, they will seem to slow down as they approach it, freezing into a fixed pattern, with the flashes coming ever less frequently and ever more redshifted. With a Kerr black hole a similar spectacle will be witnessed, except that because of frame dragging this fixed pattern will forever rotate with the spin period of the black hole. What we want to describe next is how the rotation of Kerr black holes offers a mechanism to allow energy to be drawn from them.

Recall that in relativity, mass is equivalent to energy ($E = mc^2$). A common form of energy is kinetic energy, and what the $E = mc^2$ equivalence implies is that matter can be converted into other forms of matter plus kinetic energy, and vice versa (for example in a nuclear reaction). In a black hole, all its matter-equivalent in energy is trapped, at least ignoring quantum effects which we'll get to in Chapter 7. However, rotation is a form of kinetic energy, and that can be extracted from a black hole. Note that nothing in such an extraction would be coming from inside the black hole; instead, the rotational energy in the spacetime around the black hole can be tapped. One way of doing so is called the Penrose process, named after its discoverer, the same Penrose who formulated the Cosmic Censorship Conjecture. The way it works is as follows (see figure 4.2). A space station orbiting the black hole from some distance away sends an energy mining vessel toward the black hole along a geodesic that enters the ergoregion. Equatorial geodesics are thus best for this endeavor. Once in the ergoregion, the vessel carefully aims and launches a heavy projectile at very high speed in the opposite direction to the rotation of the black hole. Of course, because of frame dragging, both the projectile and vessel will look like they're moving about the black hole in the same sense; the vessel will just be going around faster. It's important that the projectile be heavy, comparable in mass to the vessel alone, because then a large recoil will be imparted to the vessel. The projectile must be aimed so that the recoil kicks the vessel onto an orbit that sends it back out to the space station, while the projectile falls into the black hole. If shot with enough velocity, the projectile will have angular momentum with the opposite sign to that of the black hole. When the projectile is absorbed by the black hole, the black hole's spin will decrease by a corresponding amount. But total angular momentum is conserved, and therefore (Newton's third law again) the vessel must have gained this same amount of angular momentum. And this means the vessel must have gained kinetic energy.

CHAPTER FOUR

Actually, so far nothing about our discussion of the Penrose process is unusual or remarkable. In fact, if we'd run this thought experiment with the black hole replaced by the Sun, the same conservation arguments would apply. The Sun, on consuming the projectile, would have its angular momentum lessened, while the vessel gains an equivalent amount and so gains kinetic energy. However, in this case, the vessel can never gain enough kinetic energy to compensate for the energy-equivalent of the mass carried away by the projectile. For the rotating black hole something unusual is happening: If the orbit is tuned carefully and the projective aimed well, the vessel can gain so much kinetic energy that it compensates for the loss of the projectile with extra to spare. It's not easy to come up with an intuitive explanation of all that happens in the black hole case. Let us instead describe one key piece of the calculation that illustrates another bizarre property of the extreme warping of space and time around black holes and explains why it's crucial for the Penrose process that the projectile is fired from inside the ergosphere.

First, we need to make a brief digression to talk about the energy of an object in an orbit. Energy can take different forms. Rest energy is the energy of mass itself, and that is what the equation $E = mc^2$ refers to. There's also kinetic energy, which is the energy of motion. And, at least in Newtonian gravity, there's potential energy, which describes how deep in a gravitational well an object is located. Potential energy is negative because it's the energy you would have to add to an initially stationary object to lift it out of the gravitational well in which you find it. In Newtonian gravity, the total mechanical energy of an orbiting object (that is, its kinetic plus gravitational potential energy) never changes,

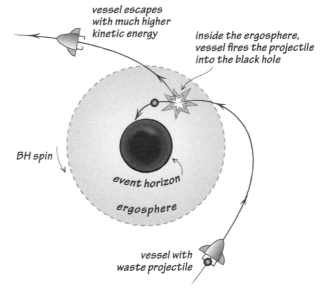

FIGURE 4.2. Illustration of the Penrose process, looking down the spin axis of the black hole onto the equatorial plane, where the mining vessel and projectile orbit.

provided the only force acting upon the object is the gravitational pull from a large, stationary mass like the sun. Any change in the kinetic energy is balanced by an equal and opposite change in the potential energy. In general relativity, it's trickier to give a definition of potential energy that makes sense in all spacetimes, but at least for an object moving in the Kerr geometry it is possible to do so. The result is consistent with the Newtonian definition far from the black hole. So the upshot is that around a Kerr black hole the total mechanical energy of an object in orbit (which now includes its rest mass energy) can be defined, and because the orbiting object follows a geodesic, this total energy never changes.

CHAPTER FOUR

Now here's where the bizarre property of frame dragging comes into play. There are geodesic orbits in the Kerr geometry, entirely confined within the ergosphere, with the property that particles following them have potential energies which are so negative that they outweigh, in magnitude, their rest mass and kinetic energies combined. That means their *total* energies are negative. This is what the Penrose process exploits: While inside the ergosphere, the mining vessel fires the projectile so that the projectile moves on one of these negative energy orbits. By conservation of energy, the mining vessel gains enough kinetic energy to make up for the rest mass equivalent in energy that left with the projectile, plus the positive equivalent of the net negative energy of the projectile. Since the projectile will subsequently be lost to the black hole, it would be a good idea to make the projectile out of waste products. Not only will the black hole consume all of it without complaint, it will give us back more energy than what we put in. Talk about green energy!

The maximum amount of energy that can be extracted from a Kerr black hole depends on how fast it is spinning. In the extremal case (the fastest the black hole can spin), about 29% of its energy is in the rotation of spacetime. This may not sound like a lot, but remember, this is rest-mass equivalent: By comparison, nuclear fission reactors harness less than one tenth of a percent of rest-mass-equivalent energy.

The geometry of spacetime inside the horizon of a rotating black hole is drastically different from Schwarzschild spacetime. Let's follow our probe across and see what happens. Initially things would seem similar to the Schwarzschild case. As before, spacetime starts to collapse, everything is dragged with it to ever smaller radii, and tidal forces start to grow. But now in Kerr, before the radius reaches zero, the collapse slows down and starts to reverse. For rapidly spinning black holes, this will happen well before tidal forces become strong enough to threaten the probe. To get some intuition for why this is happening, recall that in Newtonian mechanics, rotation is responsible for what is called the centrifugal force. This isn't a fundamental force. Instead it's a consequence of how all the constituents of matter move when set up in a spinning configuration, and how the fundamental forces within the matter have to adjust to maintain the spin. The result can be thought of as producing an effective outward force. If you've taken a sharp corner in a fast-moving car, you've felt this force. Likewise, if you've ever been on a merry-go-round you know that the faster it goes around, the tighter you have to grip the hand rails to hold on, and if you let go you'll get flung off. The analogy isn't perfect for spacetime, but it conveys the correct idea. The angular momentum in the spacetime of a Kerr black hole provides an effective centrifugal force that counters the otherwise pure gravitational attraction. As collapse inside the horizon funnels the spacetime to smaller radii, the centrifugal force gets stronger and is eventually able to counter and reverse the gravitational collapse of spacetime.

The moment the collapse has come to a halt, the probe reaches what's called the inner horizon of the black hole. Tidal forces are mild at this point, and it only takes a finite time for the probe to reach it after having crossed the event horizon. However, just because the collapse of spacetime has halted doesn't mean our problems are over and that rotation has somehow cured the singularity inside a Schwarzschild black hole. Far from it. In fact, in the mid-1960s, Roger Penrose and Stephen Hawking proved a set of singularity theorems stating that whenever a period of gravitational

CHAPTER FOUR

collapse occurs, however brief, there must be some form of singularity that eventually forms. In the Schwarzschild case, this is an all-encompassing, crushing singularity that develops over all of space within the horizon. In the Kerr solution, the nature of the singularity is quite different and rather counterintuitive, given what we know about the Schwarzschild case. When the probe reaches the inner horizon, the Kerr singularity reveals itself, but it appears in the causal *past* of the probe's worldline. It's as if the singularity had always been there, but only now is its influence reaching the probe. If this sounds bizarre, it is. And several things go wrong with the picture of spacetime that suggest this answer isn't the final one.

The first problem with the singularity showing up to the past of an observer reaching the inner horizon is that then the Einstein equations fail to uniquely predict what happens to spacetime beyond this horizon. The issue is that in a sense anything can come out of the singularity. Presumably a theory of quantum gravity will be able to tell us what actually does come out, but Einstein's equations are completely silent on this point. Out of curiosity, we'll describe below what happens if we demand that crossing the horizon spacetime is as smooth as is mathematically possible (if the metric functions are what is mathematically called "analytic"), but there is no sound physical justification for this assumption. In fact, the second problem with the inner horizon suggests exactly the opposite: Namely, in a real universe where there is matter and energy outside the black hole, spacetime becomes quite nonsmooth at the inner horizon, developing a kink-like singularity there. This is not as devastating as the infinite tidal force singularity of the Schwarzschild solution, but at the very least it throws doubt on the story that

SPINNING BLACK HOLES

the smooth, analytic extension tells. This is perhaps a good thing, because the analytic extension tells a strange story indeed.

Before getting to that strange story, first we'll explain why matter outside the black hole can so strongly affect the inner horizon. It boils down to the disparity in the flow of time inside versus outside, and how the reversal of collapse by the rotation of spacetime affects this disparity. Recall that in the Schwarzschild spacetime, this disparity is responsible for the infinite redshift and time dilation noted by outside observers, and for why they can never see anything cross the horizon. This continues to hold for Kerr, with the added twist that frame dragging introduces. In any case, outside observers can never peek past the event horizon, and so they won't be able to see the drama that unfolds at the inner horizon. The key to understanding what goes on there is to ask the opposite question: What does the *probe* see looking *back* at the outside universe as it falls toward the inner horizon? First, the time-flow effects are opposite to those seen by an outside observer looking in. The probe will notice a time contraction (i.e., events on the outside will be seen to unfold faster and faster). There will also be a gravitational blueshift, where the frequency of light from these events will get shifted to shorter wavelengths, or toward the blue end of the electromagnetic spectrum. This is similar to what the probe would have seen approaching the event horizon. Reaching the event horizon one might think that the observed time contraction and blueshift would become infinite, exactly mirroring the infinite redshift and time dilation seen by an outside observer. This is almost true for a probe that's able to use a powerful rocket engine to hover very close to the event horizon, but the experience of a probe falling freely across the event horizon is quite different. Falling across the event horizon produces a large Doppler effect that partly counteracts the gravitational time contraction, and a probe on a free-fall trajectory looking back actually won't see anything much out of the ordinary as it crosses the horizon. However, once inside, the reversal of the spacetime collapse engineered by the spin of the black hole effectively slows the probe down. As the probe reaches the inner horizon, the Doppler effect fails to counter the gravitational time flow effects, and the time contraction/blueshift does become infinite. In other words, in a finite amount of its proper time, the probe would be able to "see" the entire infinite time evolution of the outside universe! Well, not quite, which is why we put *see* in quotes. The problem here is that shorter wavelength photons have higher energy, and before reaching the inner horizon the photons would be blueshifted to such high energies that they would incinerate a probe made of any known material. This phenomenon has been referred to as the *blue sheet singularity*, and you can imagine why this would call into question assumptions about the inner horizon being smooth, except for the sterilized environment of a perfect, vacuum Kerr black hole with no photons or matter in the spacetime.

With that caveat in mind, let's explore the strange story of making the smoothest possible mathematical extension of the Kerr solution across the inner horizon. Crossing this horizon, the probe enters into a new branch of the universe. In this part of the universe, the singularity is always visible, and there is no event horizon. The singularity has the structure of a spinning ring, with curvatures and tidal forces becoming infinite as one approaches it. However, in contrast to the singularity in the Schwarzschild solution, which occurs at a moment in time in the future of all infalling paths, the Kerr ring singularity is at a definite spatial location and can be avoided by the probe. The probe can do this in a couple of ways. One is to start traveling outward to large radii again, out beyond the radial location where the inner horizon was. In this scenario, spacetime funnels the probe into a white hole region of spacetime. It is quickly ejected outward as this part of the spacetime evolves into a new Kerr black hole region, with identical spin and mass to the one whose event horizon the probe originally crossed. The probe can never return to the white hole, because as in the Schwarzschild case, the white hole is now in its past, and only the new black hole remains. However, the probe could continue this kind of motion forever, plunging into the new black hole, crossing its inner horizon, then escaping back out through the next white hole into yet another black hole region. The analytic extension of Kerr thus gives an infinite sequence of black hole universes connected by white holes.

Once it crosses the inner horizon, the other option the probe has is to continue moving inward and pass through the ring singularity. OK, big deal, it's just moved through a hoop. Couldn't it just have gone around the hoop and ended up in the same place? Astonishingly, no. The demand of maximum smoothness requires that, upon going through the ring, the probe emerges in yet another distinct region of the universe. This one can also be described by the metric of the Kerr solution with the same spin, but now minus the mass of the original Kerr black hole; in other words, the spacetime has a negative-mass naked singularity. There, the effective gravitational force produced by the singularity is actually repulsive, and geodesics "fall" away from it. Even more bizarre, there's a region of spacetime that

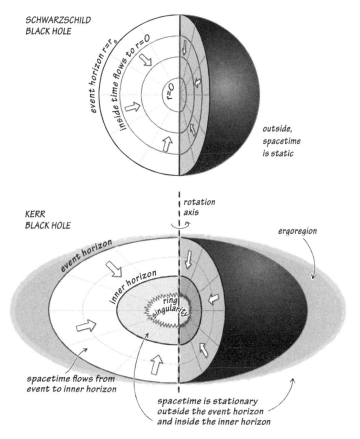

FIGURE 4.3. Schematic view of the interior structure of black holes.

has what are called *closed timelike curves*. A ring is an example of a closed curve: It has finite length, and beginning at any point on it, traveling this distance brings one back to the starting point. But "normal" closed curves are spacelike. If you travel around a ring, you're actually moving forward in time as well, and so when you get back to the starting

SPINNING BLACK HOLES

location, it's at the same spatial position, but you're in the future of when you started. This is a timelike curve, but it isn't closed. Not so with a closed timelike curve: When you get back to the starting location, you're actually back at the same spacetime event you started at.

Essentially, what we have in this region of closed timelike curves is a time machine. If you're far from the singularity, no closed timelike curves are present, and other than the repulsive nature of the singularity, spacetime would seem quite normal. However, there are trajectories you can follow (they're not geodesics, so you'd need a rocket) that would take you into the region of closed timelike curves. Once you're there, you can move in either direction of the coordinate t that measures the time of distant observers, but you'd still always be going forward in terms of your proper time. You can thus travel to any t you want and then return to the distant part of the spacetime, even arriving before you left. Of course, now all the paradoxes associated with time travel crop up: For instance, what would happen if you finished your little journey and then convinced your past self not to start it? Whether these kinds of spacetimes can exist, and how the paradoxes may be resolved if they do, are questions beyond the scope of this book. However, as with the problem of the blue sheet singularity at the inner horizon, there are hints within general relativity that regions of spacetime with closed timelike curves are unstable and can become singular if you try to put any matter or energy on one of these curves. Moreover, in a rotating black hole formed in our universe, the blue sheet singularity itself may prevent the negative mass region (and all the other Kerr universes reached through the white holes) from forming in the first place. Nevertheless it's intriguing that general relativity

CHAPTER FOUR

exhibits such strange solutions. It's easy to dismiss them as pathological, but remember how Einstein and many of his contemporaries did the same with black holes.

We finish this chapter with a brief discussion of charged black holes. We've mentioned that black holes have no "hair." In other words, they don't leave any clues in the structure of spacetime that tell you what went into them. In a sense, they have bad memories and can only remember the total mass and angular momentum of the things they consume. But what if you throw an electron into the black hole? If the black hole forgot about it, what happens to the electric charge? Wouldn't that violate charge conservation, a sacred property of particle physics? Yes it would—but thankfully black holes can have additional hair for long-range forces that relate to conserved charges. Electromagnetism describes such forces. The solutions to Maxwell's equations of electromagnetism plus the Einstein field equations that describe rotating and charged black holes are called the Kerr-Newman metrics. They are uniquely characterized by their mass, spin, and electric charge. Actually, the solution describing a *nonrotating* charged black hole was discovered many years earlier. It is called the Reissner-Nordström black hole after its discoverers. The reason the Reissner-Nordström solution was discovered much earlier is that, as in the Schwarzschild solution, the nonrotating spacetime of the Reissner-Nordström black hole is spherically symmetric, and so the field equations are mathematically much simpler. Interestingly, charge, even by itself, imbues similar properties to the interior structure of black holes as does angular momentum. There are inner horizons, blue sheet singularities, and multiple connected universes. However, without spin the ring singularity contracts to a point, and

SPINNING BLACK HOLES

so there are no negative mass regions with closed timelike curves in Reissner-Nordström spacetimes.

Another similarity to spin is that the electric field of a charged black hole produces an effective outward pressure similar to the centrifugal force in the Kerr solution. This effective pressure is related to the existence of an inner horizon in the Reissner-Nordström solution. So there's a maximum amount of charge for which the black hole becomes extremal, and beyond which the event horizon ceases to exist, revealing a naked singularity. As with spin, it seems difficult or impossible to overcharge a black hole. To do so would require adding more and more of the same charge to the black hole. But like charges repel, and eventually the repulsion would become so strong that one wouldn't be able to push in more charge. In our universe, all black holes are thought to be very close to electrically neutral. If they somehow did pick up a large quantity of charge, they'd quickly attract oppositely charged ions or electrons from the interstellar medium and be neutralized.

CHAPTER FOUR

CHAPTER FIVE

BLACK HOLES IN
THE UNIVERSE

THE 1960S AND 1970S, DUBBED THE GOLDEN AGE OF GENERAL relativity, witnessed a revolution in our understanding of black holes. The modern theoretical picture of a black hole as described in the previous chapters was largely uncovered then, using new mathematics and clever insights from many researchers, including John Wheeler, Kip Thorne, Werner Israel, Roger Penrose, and Stephen Hawking. At the same time, astronomers were peering ever deeper into the universe using more sensitive optical and radio telescopes and getting the first glimpses of what the sky looks like in X-rays. Two new, and at the time completely mysterious, classes of astronomical objects were discovered—quasars and X-ray binaries—that today we think are home to black holes.

An X-ray binary is a star system composed of an ordinary star in very close orbit with a second, unseen companion, thought to be a white dwarf, neutron star, or black hole. It is believed that matter transfer from the observed star to the unseen companion explains the intense emission of X-ray photons from these systems.

If we can't see the companion object, how do we know it's there? The answer is by the Doppler shift of photons originating in the observed star's atmosphere, caused by the orbital motion of the binary. Atoms and molecules absorb and emit photons only at distinct wavelengths, called spectral lines, and the set of lines is a unique and distinguishing feature of a particular atom or molecule. For example, sodium-vapor street lights appear bright yellow, as the predominant emission is coming from two spectral lines for sodium at 589.0 nanometers and 589.6 nanometers. When astronomers look at a star, they can see many spectral absorption and emission lines from atoms and molecules in that star's atmosphere. If the star is part of a binary, these lines will alternately redshift and blueshift in a periodic fashion because of the orbital motion of the star around its companion. This alternate redshifting and blueshifting is the same phenomenon that we discussed in connection with zoom-whirl orbits in Chapter 3.

OK, so now we know that an X-ray binary is a binary despite only seeing one star. But how do we know that in some cases, such as Cyg X-1 (a bright X-ray binary in the constellation Cygnus), the companion is a black hole? Perhaps, a skeptic might suggest, the companion is just a star that's too dim to see. The comeback to this skeptical suggestion turns out to be simple: The invisible companion is too massive to be a dim star. To flesh out this simple answer, we need to string together some additional observations, Kepler's laws of orbital motion, and the theory of stellar evolution. First, the observations. From the Doppler shift, not only can one deduce that the star is in a binary, but also detailed properties of the orbit. The period of oscillation of the spectral lines is exactly the orbital period of the binary. The precise way the Doppler shift varies over one period indicates the ellipticity of the orbit. The amplitude of the shift gives a lower limit on the maximum speed of the star. (It's the actual maximum speed only if we're viewing the orbit edge-on, but only in rare circumstances can astronomers deduce the inclination of the orbit.) Using these observations together with Kepler's laws of orbital motion gives a lower limit on the sum of the masses of the two companions in the binary. We can thus figure out the mass of the unseen companion if we can calculate the mass of the visible star. This is where the theory of stellar evolution comes in. It turns out that once we know the surface temperature and luminosity of the star (both of which may be established by direct observation), our general knowledge of stellar evolution suffices to give a fairly accurate estimate of the mass.

A star's life is driven by a competition of forces: the inward attractive force of gravity versus outward pressure forces. This statement is actually also valid for a planet like the Earth, but unlike planets a star is too massive for pressure generated by cold matter to balance gravity, at least early in its life.[1] The progenitor of a star is a collapsing cloud of gas, mostly hydrogen. As this cloud collapses, the pressure and temperature in the core increases until the temperature gets so high that nuclear fusion of hydrogen begins. Fusion releases a tremendous amount of energy in the form of photons and neutrinos, further heating the core to the point that the thermal pressure

1 You might object that the Earth is not cold; in fact its inner core is almost 6,000 kelvins. That is true; however, for the Earth, the thermal pressure is not needed to support its mass, and if we could imagine cooling it to absolute zero there would still be enough electrostatic and electron degeneracy pressure to balance gravity.

becomes sufficient to halt the collapse, and a star is born. From outward appearances the star has reached equilibrium, but the chemical composition of its core is continually evolving as it burns hydrogen fuel into helium. What happens after most of the hydrogen in the core has been depleted depends on the mass of the star. We don't want to go into all the details and possibilities here, except to note that for the most massive stars (10 to 100 times the mass of the Sun) there will be multiple equilibrium phases, punctuated by episodes of contraction that cause the core temperature and pressure to increase until a new nuclear fusion reaction begins. This process continues until an iron core forms.

Before discussing what happens in the final stages of a star's life, we can now return to the question of how knowing a star's surface luminosity and temperature tells us what its mass is. It is actually easier to think about the question going the other way: If we know the star's mass and chemical composition, we can compute the surface temperature and luminosity from the stellar structure equations. There are many technical details here, but the basic principles are as follows. A more massive star needs more thermal pressure to counter gravity, hence more nuclear fusion will take place, releasing more photons, and so the star will be brighter. The center of the star is hottest, and the temperature decreases as you move outward, reaching a minimum at the surface. What the surface temperature is depends on the structure of the star, but at least during the initial hydrogen burning phase, which astronomers call the *main sequence*, more massive stars also have higher surface temperatures. The surface temperature determines the color we perceive the star to have. Thus, from an observation of the color and brightness of a star, astronomers can work backward through these calculations and obtain an estimate of the mass and composition of a star.

Cyg X-1 contains a star with an estimated surface temperature of 30,000 kelvins and mass of 20 solar masses. Such a hot temperature makes it appear bluish in the night sky (though it's so far from the Earth that you need a good pair of binoculars or a telescope to see it), and being at least 10 times the size of the Sun classifies it as a blue supergiant star. Using this and Doppler shift observations to model the orbit, astronomers have deduced the invisible companion has a mass of approximately 15 solar masses. So why must this be a black hole? The answer again comes from the theory of stellar structure. As we explained above, as a massive star evolves, it goes through various stages of burning nuclear fuel, and the energy released provides the pressure to counteract gravity. The nuclear reactions proceed until nuclei in the mass range of iron are formed. These are the most stable nuclei, and any nuclear fusion or fission process beyond this stage requires energy.[2] At this point the atoms are completely ionized, with all the electrons floating around in a state called a *Fermi*, or *degenerate* gas. One effect of this degenerate state of matter is that it can exert considerable pressure (even at zero temperature), called degeneracy pressure. For a lower-mass star like the Sun, electron degeneracy pressure is sufficient to support the core once nuclear fusion stops (and incidentally, for lower-mass stars this will

---

2 The common form of iron, with 56 protons and neutrons, has the smallest rest mass per nucleon of any element. An isotope of nickel with 62 protons and neutrons actually has larger binding energy. A somewhat detailed account of stellar evolution is required to understand why common iron is produced more copiously than nickel-62. For our account, what matters is that the most stable nuclei are in the iron group and are the natural endpoints of fusion processes.

BLACK HOLES IN THE UNIVERSE

happen before iron is formed), and it ends its life as a white dwarf star.

For massive stars, the later stages of stellar evolution are more dramatic. Once an iron core grows to more than the so-called *Chandrasekhar limit,* which is roughly 1.4 times the mass of our Sun, electron degeneracy pressure becomes insufficient to support the core, and it collapses. The temperature and density in the core rises very rapidly, and high-energy photons begin to disintegrate the iron. In this extremely dense environment, free electrons and protons readily combine to form neutrons, and a gas of neutrons quickly forms. Neutrons are fermions, so they too exert degeneracy pressure, and it turns out to be significantly higher than electrons' degeneracy pressure—so high that it can halt the collapse of the core. This is actually a rather rapid and violent process, sending a powerful shock wave propagating outward through the star. Though the details are still a mystery, astronomers believe this is the beginning of what is eventually observed as a type II supernova. During the process, most of the outer layers of the star are blown off, but some material rains back down onto the core, which is now a proto-neutron star.

Similar to the Chandrasekhar limit for stars supported by electron degeneracy pressure, there is an upper limit to the amount of mass that neutron degeneracy pressure can support, sometimes referred to as the *Tolman-Oppenheimer-Volkoff* (TOV) limit. The physics of nuclear matter at the extreme densities thought to exist in neutron stars is not too well understood, and this translates into significant uncertainty regarding what the actual value of the TOV limit is. From observations of neutron stars we know it's at least 2 solar masses. Theory also tells us it can't be more than about 3 solar masses if we make what seems to be a reasonable assumption that sound waves in the neutron star cannot propagate faster than the speed of light. If enough material accretes onto the core to raise the mass above the TOV limit, the proto-neutron star itself collapses. There may be as yet undiscovered phases of matter beyond nuclear density, but as long as the speed of sound in these new phases is less than the speed of light, no core above 3 solar masses can be supported, and general relativity then unequivocally predicts that a black hole will form.

Let's return to Cyg X-1. We know the companion is about 15 solar masses. Visible stars exist which are even more massive (in fact, the visible star in Cyg X-1 is such a star!), but since the companion is not visible, it can't be supported by the thermal processes that hold up ordinary stars. However, 15 solar masses is way above the TOV limit. So we reason that the companion can't be an ordinary star, a white dwarf, a neutron star, or any star-like object composed of ordinary (baryonic) matter.

Could it be a "dark star" formed from dark matter? Dark matter is a hypothesized particle or family of particles that interacts very weakly (or not at all) with ordinary matter. That's why we can't "see" dark matter, as it interacts too weakly with the electromagnetic field to produce enough photons to be visible. The dark matter hypothesis arose several decades ago to explain the following observation: On galactic scales and larger, astronomers see stars and galaxies move as if there is a much stronger gravitational pull on them than can be accounted for by all known forms of surrounding matter: stars, dust, gas, light, neutrinos, and so forth. We have no idea what is causing this anomalous force, but, at least as of today, many scientists would bet it's some form of dark matter. Building on this speculation then, dark matter could clump together to form dark, compact objects, one of which could be the invisible companion in Cyg X-1. However, the dark matter hypothesis by itself does not preclude the existence of black holes (in fact, some have proposed that black holes *are* the dark matter), so one would have to layer on yet other doses of speculation to have "dark stars" be a theoretically plausible and likely answer to what's hiding in binaries like Cyg X-1.

FIGURE 5.1. What a black hole–stellar binary system, such as Cyg X-1, might look like. The star could be millions of kilometers in radius, while the black hole lurking in the center of the accretion disk has a radius of only a few hundred kilometers at most. Therefore, the inner part of the disk around the innermost stable circular orbit (where most of the X-ray emission occurs) is not resolvable on the scale of this figure. A jet of material can also be launched from the inner regions of the disk, powered by the spin of the black hole.

CHAPTER FIVE

There is additional evidence consistent with the invisible companion in Cyg X-1 being a black hole. The most convincing evidence is the bright X-ray emission coming from its vicinity. Though the visible star does emit some X-ray photons, it does not emit nearly enough to explain the observed X-ray luminosity. If the companion is a black hole, it is close enough to the star that it captures a large amount of gas and dust from the stellar wind. This material orbits the black hole in a puffy disk, but due to the viscosity of the material and magnetic field effects, it slowly migrates toward the black hole until it reaches the innermost stable circular orbit (ISCO). Recall from Chapter 3 that the ISCO is the closest any particle following a geodesic can circle around a black hole without falling in. For a nonrotating Schwarzschild black hole, the ISCO is 3 times the event horizon radius, but it moves closer in for a spinning black hole, eventually hugging the event horizon for a maximally rotating Kerr black hole. After reaching the ISCO, the gas quickly plunges into the black hole. The black hole is thus continually accreting material, and the disk of matter about the black hole is referred to as an accretion disk. During the long inward migration to the ISCO, the gas heats up. The source of energy for the heating comes from the gravitational potential energy released as the gas moves closer to the black hole.[3] The closer the gas moves to the black hole,

---

3 This is the same potential energy that entered our discussion of orbits in Chapter 4. The difference here is that the kinetic energy that the gas molecules gain to account for the decrease in their potential energy as they orbit closer to the black hole is evenly distributed throughout the gas via collisions between nearby molecules. This process eventually registers as a corresponding increase in the temperature of the gas. This gravitational potential energy is also the same type of energy that we associate with objects at different heights relative to the ground on Earth; we discuss this and the connection to accretion disks in a bit more detail later on.

the hotter it gets, which means the corresponding photons that are emitted have higher average energy. The highest-energy photons thus come from the vicinity of the ISCO. The size of the ISCO is related to the mass of the black hole, and thus the highest-energy photons emitted from an accretion disk are an indication of the size of the black hole. For black holes several times the mass of the Sun, like the one we think exists in Cyg X-1, this corresponds to X-ray emission. Moreover, irregular flow of material in the disk can result in variations in the brightness of the X-rays, called *quasi-periodic oscillations*, and the shortest timescale of these oscillations corresponds to the orbital period of particles at the ISCO. For stellar mass black holes, this variability is on the order of a few hundred hertz (meaning, a timescale of variation of several milliseconds), which is observed in Cyg X-1 and many other candidate X-ray binary black hole systems.

As simple as the answer is to why the companion is a black hole (it's too massive not to be one), as we've explained, it relies on a long chain of theoretical arguments. Some of these arguments are quite well supported by observation and experiment (stellar evolution below nuclear densities), others are a bit uncertain (the nature of matter at nuclear densities), and one is highly plausible but entirely speculative (that there aren't massive, compact dark matter "stars" that shine in X-rays). So a more conservative statement would be that the observed properties of high-mass X-ray binaries like Cyg X-1 are consistent with the black hole interpretation, and that no one has yet come up with an alternative explanation within conventional, well-tested theories. Before September 14, 2015, this was about as good an argument for the physical reality of black holes as we

CHAPTER FIVE

could make. On that day, the LIGO detection of the merger of two black holes changed everything. Science can never make a 100% definitive statement on something like this, but observing a collision in gravitational waves cleanly cuts out all the nongravitational theoretical arguments that are needed to support the Cyg X-1 case (or that of quasars, which we discuss next), only relying on the properties of vacuum general relativity. We will give a fuller account of this exciting detection—which really amounts to the founding of a new branch of astronomy—in Chapter 6.

There is a second population of black holes in the universe for which evidence has steadily been accumulating since the late 1960s, namely the black holes first associated with quasars. The name "quasar" was coined in the 1960s and derives from the term "quasi-stellar object," which at the time translated as: "We don't know what these things are, but they sure are bright." We now think that quasars are examples of *active galactic nuclei* (AGN), where a small region in the center of a galaxy surrounding a huge black hole is filled with matter that emits copious radiation as it spirals into the black hole. Quasars are extremely bright and extremely distant: many billions of light-years away. To understand just how distant that is, consider that our galaxy is about 100,000 light-years across! (One light-year equals slightly less than 10 trillion kilometers.) Quasars are far brighter than a whole galaxy filled with billions of stars. The black holes at the heart of quasars are thought to have masses ranging from several million to a few billion times the mass of the Sun, and they are referred to as *supermassive* black holes. Thus, in a sense, quasars are far more awesome than even the first black hole merger detected by LIGO, which threw off a mere 3 solar masses' worth of energy from black holes whose combined mass was roughly 65 solar masses. However, those 3 solar masses' worth of energy blasted out of the black hole merger in a few tenths of a second. Imagine what might happen if two supermassive black holes merged!

As with X-ray binaries, we should be circumspect about how confidently we can claim that quasars are, at their heart, black holes. This claim goes back to 1969, when British astrophysicist Donald Lynden Bell realized that one way to explain the luminosity of AGN was if they were powered by black holes, though he amusingly used the term "Schwarzschild throat" in place of "black hole," a term which John Wheeler had suggested only a few years previously. The mechanism Lynden Bell suggested to explain a quasar's emission is the same as the source of Cyg X-1's X-ray emission: an accretion disk. The difference for AGN is that since the black holes are so much larger, the luminosity from the accretion disk peaks at correspondingly longer wavelengths, and so they are brightest in the radio to optical part of the spectrum. This larger size also explains variability on time-scales of order minutes to hours seen in the emission from quasars: This is the equivalent of the several millisecond quasi-period oscillations seen in Cyg X-1's emission, but scaled up to the longer periods associated with the larger ISCOs of the supermassive black holes. What feeds the accretion disk of a supermassive black hole is gas and dust falling in from the surrounding galaxy, as well as the occasional star that wanders too close to the black hole and gets torn apart by the strong tidal forces near the black hole's horizon. Overall, the matter consumed by the black hole from its accretion disk can amount to tens or even hundreds of solar masses every year. The accretion disks, rather than the black holes themselves, produce light. They are bright beacons of young galaxies, and the light that we see from them today was typically produced billions of years ago.

At first glance it may seem surprising that an accretion disk can be a sufficiently powerful source of energy to make a quasar outshine all the light from the stars in a galaxy. The source of this energy is the gravitational potential energy of the material orbiting the black hole. This is the same gravitational potential energy we deal with here on Earth every day. For example, gravitational potential energy is what we are tapping in hydroelectric power plants. Water running down from a higher elevation to a lower elevation is giving up gravitational potential energy, which the power plant converts to electricity that we could use to power lights. Quasars are similar, but they produce something like a million trillion trillion times as much power as a large hydroelectric plant. When dealing with black holes, a useful way to characterize the amount of potential energy that can be converted to other forms when falling in from large distances down to the ISCO is as a fraction of the total rest-mass energy ($E = mc^2$) of the matter. This number depends on the spin of the black hole, as the location of the ISCO depends on these factors. The percentage is 6% for a non-spinning black hole, and increases to 42% for a maximally spinning black hole.[4] These are huge numbers. For comparison, the amount of energy available from water that runs

4 Note that this gravitational potential energy is a different source of energy than that available in the rotation of the black hole that the Penrose process taps. For an accretion disk, the reason a rotating black hole gives a higher percentage is that spin moves the ISCO closer to the horizon, so there's more distance to extract potential energy from the gas as it migrates in. Once it reaches the ISCO, matter plunges into the black hole too quickly for the kinetic energy it gains to be transferred to heating the surrounding gas.

down a hill 100 meters high is a trillionth of a percent of the rest-mass energy.[5] The most efficient source of energy at our disposal today is nuclear energy in uranium fission reactors. If all the uranium fuel is used up in a reactor, a rest-mass equivalent of just under 0.1% energy is released. Still a small fraction compared to what is theoretically possible in a black hole powered accretion disk. It is thought that most AGN operate close to, but not at, the maximum possible efficiency. The main reason is that as the gas heats up and starts radiating copious amounts of energy, the thermal pressure becomes large enough to counter the inward flow of the gas, and some of it gets blown off by the analog of wind.

As the notion of a black hole took hold, and astronomers started to accept that black holes could explain the nature of quasars, a natural question arose as to whether galaxies that do not have active galactic nuclei might still harbor supermassive black holes. This possibility was in fact suggested by Lynden Bell in his 1969 paper. These would be dormant in the sense that they do not have much gas bound in an accretion disk, and so they would not be as luminous. For nearby galaxies, measurements of the average Doppler shifts from the collection of stars near the core of the galaxy can be made. The inferred orbital dynamics implies that there are in fact central supermassive black holes in all larger galaxies. This is certainly true for our Milky Way, the center of which is close enough that several individual stellar orbits about it can be resolved. The stars orbit what is apparently a black hole roughly 4 million times the mass

---

5 However, if we do a thought experiment collapsing the Earth to a black hole, its ISCO would be a few centimeters from its center, and letting water run down to *that* radius from 100 meters up would give similarly huge numbers.

CHAPTER FIVE

of the Sun. On the scale of supermassive black holes this is on the lower end, but not inconsistent with the size of the Milky Way (larger galaxies tend to have larger black holes). The position of the black hole is coincident with a bright radio source in the constellation Sagittarius, called Sagittarius A⋆ (or simply Sgr A⋆). The emission from Sgr A⋆ is believed to come from the accretion disk around the black hole, but compared to an AGN, Sgr A⋆ is extremely dim and definitely in a dormant state.

Unlike their stellar mass counterparts, there is no universally accepted theory of how supermassive black holes originally formed. One possibility is they were seeded by the collapse of the first generation of massive stars formed a few hundred million years after the Big Bang (which occurred almost 14 billion years ago). These black holes would initially have been ten to a hundred times the mass of the Sun, but after settling to the centers of newly formed galaxies, they would grow by gas accretion and mergers with other black holes. A challenge for this hypothesis is to explain the observation of some very distant quasars, whose light reaching us today has come from only about a billion years after the Big Bang. This implies that a supermassive black hole is already there, and the challenge for the accretion/merger hypothesis is to explain how so much growth could have occurred in what is cosmologically a short time of a few hundred million years. Another hypothesis suggests that the seeds of present-day supermassive black holes came from a time much earlier in the universe (or even at or before what we call the Big Bang). This putative class of black holes are called primordial black holes. At present there are no entirely convincing theoretical mechanisms for their formation, nor does other observational evidence exist for them.

BLACK HOLES IN THE UNIVERSE

*a disk around a spherical object
in Newtonian gravity (e.g., Saturn)*

about a black hole (e.g., "Gargantua")

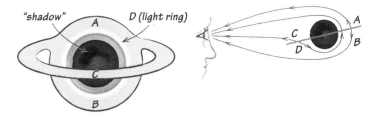

FIGURE 5.2. The "shadow" of a black hole. In Newtonian gravity (top), light coming from a disk orbiting a massive object is not bent, and we would see an undistorted image of the unobscured part of the disk. Around a black hole (bottom), the curvature of spacetime is so large that light paths are severely bent, so much so that all parts of the disk are visible. The paths of a few of these light rays are illustrated on the bottom right, giving rise to the image shown on the bottom left.

We conclude this chapter by briefly mentioning an exciting astronomical project called the Event Horizon Telescope that is already beginning to provide images of the so-called shadow of our own galactic black hole, and the billion-plus solar mass monster in the galaxy M87, which is relatively close to us, at just over 50 million light-years away. What's special about these two black holes is that their Schwarzschild radii have the largest angular sizes on the sky of any known black hole. For Sgr A* that's because it's (relatively speaking) so close to us, and for M87 because it's so huge. The Event Horizon Telescope is actually a col-lection of radio telescopes at widely spread-out locations across the globe that work in tandem. This allows them to exploit a parallax-like effect when simultaneously measur-ing the radio waves from the same source—a technique called interferometry—to effectively achieve the angular resolving power that a telescope the size of the Earth would have. As a result, they can resolve very small-scale features on the sky, which is needed to see close to the horizons of these black holes. For example, the angular extent of the event horizon of Sgr A★ is only 6 nanodegrees. Trying to resolve that is like trying to make out details on the face of a silver dollar lying on the surface of the Moon! This inter-ferometry technique cannot give the telescopes the same light-gathering power as a hypothetical telescope the size of the Earth (they still can only gather as much light as falls on their combined surface area), but at least for the purpose of observing Sgr A★ and M87, resolving power is much more important. Of course, they won't be able to "see" either of these black holes themselves, but instead they will see light coming from the accretion disks whirling around them. This light will (for the most part) follow the geodesics of the black hole spacetime, but as we saw in Chapters 3 and 4, very near the horizon the warping of spacetime is so strong that the photon trajectories will be extremely curved, with some even traveling a few times around the black hole near its light ring before escaping to us. The effect is that the accretion disk will look quite warped. An inner, circular shaped part of the image of the accretion disk, correspond-ing to a region a few times the size of the Schwarzschild radius, will appear darker (the shadow), except for a bright ring marking the location of the light ring. If we're viewing the accretion disk edge-on, the nearer part of the accretion disk will cut across this shadow. Also, above and below the shadow, we'll actually be able to see part of the disk that's *behind* the black hole, again because of the bending of the photons as they stream away.

CHAPTER FIVE

CHAPTER SIX

BLACK HOLE COLLISIONS

In Chapters 3–5, we focused on black holes in isolation from one another. We were (and are!) very interested in the way stars orbit around supermassive black holes, and in how matter forms accretion disks around black holes, because these phenomena provide the best evidence we have for the existence of black holes—that is, the best until the LIGO observation of gravitational waves from a collision of two black holes. That collision took place more than a billion years ago and a comparable number of light-years away.[1] In this chapter we will explain some of the theory behind this remarkable event: what gravitational waves are, why black holes collide and in doing so produce gravitational waves, and why it took 100 years from when Einstein

---

1 Because of the expansion of the universe, the distance between us and the site of the black hole collision is now somewhat greater than the distance traveled by the gravitational waves on their way to us.

published the theory of general relativity before scientists were able to directly measure gravitational waves for the first time.

Black hole collisions are the most violent events that are possible in general relativity. The Big Bang at the start of the universe was undoubtedly still more dramatic, but a larger theory than general relativity is required to describe the beginning of time itself. Physicists are still searching for the right theoretical framework in which to give a full account of the Big Bang. Black hole collisions do not require any such larger theory: By all accounts, just the Einstein equations, $G_{\mu\nu} = 8\pi G_N T_{\mu\nu}/c^4$, should be enough. In fact, for many black hole collisions, we can probably ignore the stress-energy tensor ($T_{\mu\nu}$, which would be zero if there is no matter present), because the total energy in matter outside the black holes is small compared to the rest energy of the black holes themselves. Thus, to describe black hole collisions, we are called on to solve a supremely simple-looking set of equations: $G_{\mu\nu} = 0$. A famous picture of Einstein shows him in the act of writing down an equivalent set of equations: $R_{\mu\nu} = 0$. Here $R_{\mu\nu}$ is the so-called Ricci tensor, closely related to the Einstein tensor and essentially equivalent to it in the absence of matter. Einstein's choice of subscripted indices i and k is a matter only of taste; he could just as well have written $R_{\mu\nu} = 0$.

We've discussed the Einstein field equations before, but as we launch into an account of black hole collisions it's worth reviewing the intuition of what the field equations describe. Briefly, the field equations give mathematical form to the idea that matter tells spacetime how to curve. What do the field equations permit spacetime to do when there is no matter? One example of a permitted spacetime is to have

FIGURE 6.1. Einstein pictured with the vacuum field equations of general relativity, which are a special case of the Einstein equations in the absence of matter.

no curvature at all. In other words, completely flat spacetime is a solution of the vacuum Einstein equations—but it is not the only one. Indeed, another example of a spacetime permitted by the vacuum field equations is an isolated black hole. Inside the black hole horizon, as we have learned, there could be singularities or other peculiar features that we might associate with a nonzero stress-energy tensor. But outside the horizon, it's possible for there to be no matter at all. We shouldn't feel too committed to any account of the inside of the black hole, because no signals from inside can possibly reach us. So the most economical view is that an isolated black hole is an example of spacetime curving even in the absence of matter. Black holes moving around one another provide yet another example of a solution of

BLACK HOLE COLLISIONS

the vacuum field equations. Eventually, a pair of black holes orbiting each other will spiral in and merge into a rapidly spinning Kerr black hole. This is the sort of event that was observed by LIGO on September 14, 2015.

One more important class of solutions of the vacuum field equations is gravitational wave geometries. As we explained in Chapter 1, we should understand gravitational waves in terms similar to the way Maxwell described light. Recall that light is a traveling wave of electric and magnetic fields, arranged so that the spatial variation of the electric field causes the time variation of the magnetic field, and vice versa, all according to Maxwell's equations of electromagnetism. Usually we think of electric fields as arising from the presence of electric charges, while magnetic fields result from the presence of electric currents; but in light, the electric and magnetic fields, once created, keep propagating forever—or at least until they encounter some matter which absorbs or scatters them. Gravitational waves are similar: A disturbance of the flat spacetime metric propagates forever, with spatial variations of the metric causing timelike variations according to the vacuum field equations.

Let's pursue the analogy between gravitational waves and light a little further. Electromagnetic waves are produced by accelerating electric charges. For example, a radio tower works by sending high-frequency alternating currents back and forth along conductors. These currents—which are nothing more nor less than electric charges which accelerate back and forth—are the first cause of electric and magnetic fields which then propagate outward to be picked up by a radio that one of us might operate. Radio waves are indeed the same thing as light, just with longer wavelengths, and visible light can in principle be produced through similar

CHAPTER SIX

back-and-forth acceleration of charges. Likewise, gravitational waves are produced by accelerating matter. A common form of acceleration in gravitational systems is the centripetal acceleration of circular orbits. An example which we summarized at the end of Chapter 2 is that binary star systems produce gravitational radiation because of the orbital motion of the two stars around each other. The energy carried off by this radiation leads to a perceptible inspiral of the stars' orbits. Perhaps it's not too surprising, then, that black holes spiraling in toward each other should also emit gravitational radiation. However, from a philosophical point of view, it's very surprising that gravitational radiation can emerge from a system which is nothing but empty space (by "empty" we mean that it solves the vacuum field equations). This brings home a point that we've made before: Gravity itself gravitates.

The gravitational waves detected by LIGO have been compared to sound; perhaps most famously, leaders of LIGO have referred to the chirps and thumps they hear as "the music of the cosmos." This analogy is both inspiring and useful, but here we want to emphasize how different gravitational waves really are from sound. Sound is a compressive wave in air. What that means is that a sound wave consists of alternating regions of high pressure and low pressure propagating through a volume of air. Individual air molecules are constantly moving around in a chaotic thermal dance, but on top of that complicated, random motion, sound causes molecules, on average, to sweep a little bit toward the listener as a high pressure region presses them forward, then a little bit away from the listener as a low pressure region sucks them back. This is an example of a longitudinal wave, where the word "longitudinal" refers to the fact that the internal motion comprising the wave is back and forth along the same axis that defines the direction of propagation of the wave. In contrast, an everyday example of a transverse wave is a wave on a taut string stretched out in a horizontal direction. If we jerk one end up and down, we can then watch the up-down disturbance move along the string. "Transverse" refers to the way the internal motion comprising the wave (up and down in the example we've described) is at right angles to the overall motion of the wave (horizontal in this case). Gravitational waves (and, incidentally, also light) are transverse waves. An interesting implication is that an explosion where all matter accelerates outward in a perfectly spherical shell produces no gravitational radiation at all.[2] Trying to produce gravitational waves in this manner would be like trying to create an up-down transverse oscillation of a taut string just by pulling the string tighter without moving it up or down at all. Sound behaves very differently: An exploding shell of matter would make a splendidly loud sound, precisely because the explosion constitutes motion in the same outward direction in which the sound naturally propagates.

An even more basic difference between sound and gravitational waves is that sound requires a medium to propagate. Usually that medium is air, but sound can also travel through water or through solid materials. However, sound cannot travel through a vacuum. Light can travel through a vacuum, and so can gravitational waves. According to modern thinking, spacetime is the medium of gravitational waves in a manner closely analogous to how matter provides

2 Certain supernovae may be examples in nature of explosions that occur in a nearly spherical shell, or at least close enough to spherical that despite being extremely violent accelerations of matter they won't produce much in the way of gravitational waves.

CHAPTER SIX

a medium for sound waves. From this point of view, it is really the transverse character of gravitational waves that distinguishes them from sound, not the presence or absence of a material medium per se.

The transverse character of gravitational waves is crucially important to the way gravitational wave detectors are designed. Let us therefore consider a bit more carefully what exactly a gravitational wave looks like. For the sake of visualization, imagine a gravitational wave propagating vertically downward toward the LIGO detector in Livingston, Louisiana. A gravitational wave is nothing but a disturbance of the spacetime metric, so all it can do is change distances. To understand exactly how it does so, imagine that in place of the LIGO detector, you set up a three-dimensional cubic grid of measurement devices, all with synchronized clocks, so that by exchanging light rays they can track how the spatial distances between them change over time. (Perhaps LIGO scientists would have done exactly this if their budget had permitted!) When gravitational waves are absent, the configuration of devices remains stationary. What happens when the gravitational wave hits? The first point to appreciate is that vertical distances do not change at all. That's because gravitational waves are transverse, and the gravitational wave we have in mind is coming vertically downward. However, in the north-south horizontal direction, the distances between the devices will at first increase to some maximum separation, then decrease to a minimum separation, and so on as each cycle of the wave passes. In the east-west direction, the same changes in distances will be observed, but exactly out of phase with the north-south distances. In other words, the gravitational wave is simultaneously "stretching" space in the north-south direction

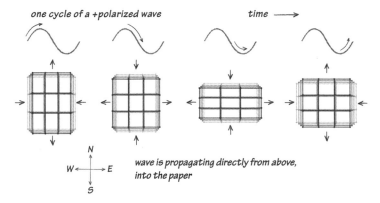

FIGURE 6.2. Effect of a passing gravitational wave on a cubic grid. We can imagine detectors at each point on the grid which measure how distances along the grid change with time.

and "squeezing" space in the east-west direction, and then vice versa.

The LIGO detector at Livingston is much simpler than the cubic grid of measurement devices that we imagine here. It has one arm that runs 4 kilometers a few degrees east of south from the central facility, and one that runs 4 kilometers at right angles to the first arm, a few degrees south of west. The precise directions of the arms don't matter for our purposes, so for the remainder of our discussion we'll suppose that they run due south and due west. It is not too much of an idealization to say that LIGO is effectively like three measurement devices of the sort imagined in the previous paragraph: one at the central facility and one at the end of each arm. All the fancy laser interferometry that makes LIGO work can be idealized as the way in which these three measurement devices exchange light rays to track how the distances between them change over time.

CHAPTER SIX

Actually, LIGO doesn't track absolute distances; instead, it tracks how the *difference* of the distances along its two arms changes with time. In short, LIGO measures much less about spacetime than our imagined cubic lattice of measurement devices, but it measures just enough to detect the stretch-squeeze pattern of a gravitational wave of the type described in the previous paragraph.

Now suppose that a gravitational wave comes along that stretches spacetime along a northwest-southeast axis while squeezing it along a northeast-southwest axis. It stands to reason that this type of gravitational wave must be just as common as the one we were focusing on before. We can describe the stretch-squeeze pattern aligned along north-south and east-west directions as the plus-polarized wave, and then the pattern aligned along northwest-southeast and northeast-southwest axes would be described as cross-polarized. Those names come from the resemblance of the stretch-squeeze patterns to the symbols + and × respectively. Said another way, the pattern of a plus-polarized wave is a 45° rotated version of a cross-polarized wave.

Now, here's the shocking fact. LIGO Livingston is blind to cross-polarized gravitational waves! That's because the cross pattern doesn't change the difference in distance along the two arms. What happens instead when the cross pattern hits the detector is that the angle between the arms first increases and decreases—by an imperceptibly small amount. Fortunately, most gravitational radiation is neither purely cross-polarized nor purely plus-polarized, but instead some mix of the two. So LIGO Livingston's sensitivity to only one of the two possible polarizations is not nearly as much of a liability as it first appears. Recall as well that we have discussed only gravitational waves coming directly from above;

BLACK HOLE COLLISIONS

but surely gravitational waves travel in every direction more or less democratically. LIGO Livingston's sensitivity to gravitational waves in fact varies both with direction and polarization—and the same is true for the LIGO detector in Hanford, Washington. The situation is not too different from old-fashioned rabbit-ear TV antennas, which sometimes call for delicate adjustment to pick up the best signal.

In the end, what each of the two LIGO detectors measures is very simple: the difference in distance along its two arms. But that measurement is performed with astonishing accuracy. For example, when LIGO reaches final design sensitivity (expected around 2018–2020), it will be able to measure changes in distance as small as 10^{-19} meters. That's one ten-thousandth the size of a proton! Such incredible accuracy is necessary because the stretching and squeezing of spacetime by gravitational waves is exceedingly small. For example, the orbital motion of Mercury around the Sun creates gravitational waves, but LIGO cannot measure them, both because they are too weak and because they are at far too low a frequency for LIGO to be sensitive to them. Before September 14, 2015, no measurement device was sensitive enough to notice any gravitational waves at all. The path to the first detection was long and arduous, with many scientists across the world having spent decades building ever more sensitive gravitational wave detectors. Now, at the dawn of the era of gravitational wave astronomy, LIGO can only detect cataclysmic events like black hole mergers. It is hoped that as the sensitivity of gravitational wave detectors improve, we will eventually see more subtle signals, such as the gravitational waves from neutron star collisions. Gravity thus provides us a study in contradictions: It is the one force strong enough to overwhelm all others and cause black hole formation, but

CHAPTER SIX

at the same time it is so weak that gravitational echoes even of terrible events like the collision of neutron stars are still imperceptible to our most sensitive measurement devices.

Let's pause to consider what we've learned so far about black hole collisions and their detection. The subject is a simple one in the sense that "all" we are doing is studying solutions to the vacuum Einstein equations, $G_{\mu\nu} = 0$. The trouble, as we'll explain below, is that these equations are extremely hard to solve in practice. The solutions we are interested in describe black holes spiraling into one another and emitting gravitational radiation in the process. This gravitational radiation propagates through spacetime to be detected by LIGO as a deformation of spacetime in which distances are briefly compressed in one direction while being stretched in an orthogonal direction, and then stretched in the first direction while being compressed in the second. What we want to do next is to give a fuller account of what is going on in black hole collisions and how their description in general relativity is translated into the practical methods used at LIGO to search for gravitational wave events.

On first hearing the term "black hole collision," one might naturally call to mind two black holes zooming toward each other and colliding head on. This could happen, and it would generate a lot of gravitational radiation, but it's thought to be a very rare event. The reason is that black holes aren't very common in the universe (lucky for us!), and there's a lot of space between them compared to their size. Even in a crowded environment, like a globular cluster whose core may contain several hundred black holes with an average nearest-neighbor separation as small as a light–month, such a collision might happen only once every billion years or more. This is because it is just very unusual for two black holes to find themselves on a collision course by chance.

What happens more commonly is that both stars in a stellar binary system are massive enough so that at the end of their lives they collapse to black holes, forming a black hole binary. These black holes won't immediately collide, but they are fated to do so eventually because they do not have sufficient velocity to escape each other's pull. At first, they'll simply be in orbit around each other. Suppose for the sake of a definite discussion that the two black holes in question are similar to the ones that created the signal observed by LIGO on September 14, 2015. We'll simplify our account slightly by taking each black hole to have 32 solar masses, and we'll suppose that their initial separation is 384,000 kilometers, which is the average distance between the Earth and the Moon. Let's also assume that neither black hole has appreciable spin, so that when they are far apart, each one is well described by the Schwarzschild solution. Each of them then has a spherical event horizon of radius 95 kilometers. Then the slow inspiral of their orbits, due to energy lost to gravitational radiation, takes about 210 years, ending when the horizons touch. The farther apart the black holes were initially, the longer the inspiral would have taken. In fact, the time scales with the fourth power of the initial separation. In other words, if the same two black holes had initially been twice as far apart, it would have taken them 16 times longer to complete their inspiral. This scaling makes precise the statement that the inspiral starts out slowly and becomes faster and faster as the black holes approach each other. Indeed, the early stages of the inspiral of the black holes first detected by LIGO may have lasted billions of years. As we'll soon discuss, the final phase that LIGO was sensitive to only lasted a few milliseconds.

CHAPTER SIX

The frequency of the gravitational waves emitted by the binary black hole system we're discussing is twice the orbital frequency. This frequency starts out small, and it increases as the inspiral progresses, reflecting the fact that the black holes revolve around each other faster and faster as they get closer. It seems counterintuitive that a loss of energy to gravitational radiation should cause progressively faster orbital motion. It happens that way because of the balance of potential and kinetic energy: As the black holes draw closer, their gravitational potential energy drops so quickly that they are able to increase their kinetic energy while at the same time emitting gravitational waves.

The gradually increasing frequency is important to the way LIGO searches for black hole collisions. LIGO is sensitive to gravitational wave frequencies in the range of 30–1,000 hertz. For sound waves, this is in the range of human hearing, and so, notwithstanding our discussion of transverse versus longitudinal waves, it is fitting that LIGO scientists describe themselves as listening for the gravitational wave sounds that the universe makes. The inspiral sound of the binary system we have described above reaches 30 hertz (a deep, rumbling bass) at a separation of 990 km, and this is only 290 milliseconds before merger. At this stage, the black holes are whirling about each other at speeds of 47,000 kilometers per second, a little over 15% of the speed of light. The frequency rapidly increases, and merger begins near 190 hertz (close to a G below middle C, in the frequency range of a normal speaking voice). At this moment the two black holes are flying at speeds near 86,000 kilometers per second—almost one third the speed of light. The event horizons now coalesce into a single structure, resembling in shape a rotating peanut shell.

BLACK HOLE COLLISIONS

It might occur to the reader that the frequencies under discussion—tens to hundreds of hertz—are only modestly smaller than the quasiperiodic oscillation frequencies mentioned in connection with Cyg X-1. Is there a connection? Indeed there is! Recall that the few-hundred-hertz frequency range, corresponding to a timescale of a few milliseconds, characterizes the shortest timescale of variability of X-ray emissions from the accretion disk in Cyg X-1, and the explanation was that this timescale should correspond to the orbital period of particles at the ISCO of the black hole that forms the heart of Cyg X-1. Similarly, a modestly smaller frequency range, corresponding to a modestly larger timescale, characterizes the frantic whirl of a pair of 32-solar-mass black holes on the point of merging.

Once the event horizons merge, we intuitively know what to expect based on the no-hair theorems: The highly irregular, peanut-shaped black hole should settle down to the slightly flattened spherelike shape of a Kerr black hole. This process is called ringdown, and it is initially quite violent. Gravitational waves of many different frequencies are emitted during ringdown, but the strongest has a frequency of 300 hertz (a D above middle C). This ringdown wave rapidly decays: Its amplitude of vibration drops by a factor of 10 every 8.6 milliseconds. So 8.6 milliseconds after merger, it is 10 times smaller than at merger; 17 milliseconds after merger, it is 100 times smaller; 26 milliseconds after merger, it is 1,000 times smaller; and so on. In a fraction of a second, then, the merger remnant settles down to a perfectly quiescent Kerr black hole.

To recap: Gravitational waves are produced throughout the evolution of a black hole binary through a long, gradual inspiral, ending when the black hole horizons merge and

CHAPTER SIX

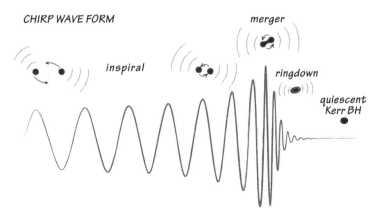

FIGURE 6.3. The "chirp" gravitational waveform from two colliding black holes.

ring down into a stationary Kerr solution. By far the most powerful gravitational radiation is emitted during the few milliseconds surrounding the merger event itself. The frequency of this radiation is in the audible range, ramping up from a low rumble to a final loud "whoop." The waveform as a whole is what we call the gravitational wave chirp, and in the audible region to which LIGO is mainly sensitive it lasts for only a fraction of a second. Collisions of larger black holes would give a lower-pitched chirp—in fact, mergers of black holes with much more than 60 solar masses would chirp at too low a frequency for LIGO to hear. Conversely, a less massive merger would chirp for longer in the frequency range of LIGO's sensitivity, and it would end with a higher-pitched whoop. The pitch of the chirp is related to the total mass of the merging black holes because the chirp comes from the last few whirling orbits before the merger, and the duration of these orbits is proportional to the final horizon radius, which in turn is proportional to the total mass.

BLACK HOLE COLLISIONS

The first event LIGO heard happened more than a billion years ago. That's long enough ago that the entire waveform was redshifted by almost 10% due to cosmic expansion— an effect similar to Doppler redshifting. In other words, by the time it reached the Earth, the chirp sounded like that of black holes 10% heavier than what they actually were. How, you might wonder, can we tell the difference between a 10% redshifted chirp coming from distant black holes and an un-redshifted chirp from nearby black holes that are 10% more massive? The answer is that the amplitude of the signal from nearby black holes would be much larger. Amplitude falls off as the inverse of distance. So if we understand from first principles (that is, by solving Einstein's vacuum field equations) the strength and frequency of the chirp produced by any given black hole merger, we can judge from the strength and frequency of an observed chirp both the distance to the black hole merger that produced it and the total mass involved.

Despite tremendous loss of energy to gravitational waves, the final remnant of a black hole merger winds up with quite a lot of rotational energy in the form of spin. Exactly how fast the remnant spins also depends on what the spins of the two initial black holes were. In the example we've discussed, they were both initially not spinning, and then the new Kerr black hole has a spin equal to 70% of the maximum allowed by general relativity. Its mass is 61 solar masses, with the gravitational waves having carried away the equivalent of 3 solar masses of energy. Total energy is conserved when we use $E = mc^2$: We started with $32 + 32 = 64$ solar masses, and we ended up with 61 solar masses in the remnant plus 3 solar masses' worth of gravitational radiation. Thus about 5% of the total mass of the binary emerged as gravitational radiation, with the vast majority of the radiation streaming out

CHAPTER SIX

during the last several orbits, the merger, and the ringdown. As a percentage, 5% energy emission doesn't sound too impressive. However, the power (that is, the rate of energy emission) is almost unimaginable. This 3-solar-mass equivalent in energy is radiated away in a fraction of a second, reaching a peak luminosity of 4×10^{49} watts. It's difficult to fathom how huge such a number is, but consider the following. The Sun has a luminosity of 4×10^{26} watts, which is roughly 20 trillion times the rate of energy consumption of all human activities combined. There are about 100 billion stars in our galaxy, and about 100 billion galaxies in the observable universe. As a rough approximation, each star is on average as luminous as the Sun, so the luminosity in starlight of the entire universe is about 10,000 billion billion Suns' worth. In power, that still only comes to a tenth of the luminosity of the last few milliseconds of this single black hole collision! That is the kind of cataclysmic event required to produce spacetime ripples large enough so that we can measure them here on Earth.

Won't such an energy release rip the fabric of spacetime to shreds? We're being a bit facetious with our language here; a more serious way to pose the question would be to wonder whether excitation of such extremely powerful vibrations in the geometry of space and time could lead to new singularities, whether clothed or naked, outside those already present in the black holes that collided. The answer is no: As strong as these gravitational waves are, they are still not quite powerful enough to do that. If the luminosity had approached 4×10^{52} watts—the so-called Planck luminosity, obtained by combining Newton's gravitational constant and the speed of light into a quantity with units of power—the answer might have been different.

BLACK HOLE COLLISIONS

So how large are the spacetime vibrations caused by this merger? Very close to the binary it would be difficult to characterize what aspect of the geometry could be attributed to the gravitational wave versus what could simply be associated with the motion of the two black holes. At distances greater than about 10 times the radius of the final orbit, it becomes possible to clearly discern the plus- and cross-polarized gravitational waves described above—and both kinds are indeed present. Let's choose a point 5,000 kilometers away from the merger; that's about 50 times the radius of the final orbit. The maximum fractional amount of stretching and squeezing at that distance will be about 0.3%. For example, suppose 6-foot-tall Alice was there observing the event. She would get stretched and squeezed by about a fifth of an inch (1/2 cm) head-to-toe. That amount of stretching and squeezing would be easy to measure (though it might be uncomfortable). Not so for LIGO here on Earth, which is roughly a billion light-years away. Because of this vastly greater distance, the amplitude would be smaller than what Alice could measure by a factor of 2×10^{18}. That's why LIGO was designed to have the seemingly ridiculous sensitivity of being able to measure a change of 1/10,000 the size of a proton over a 4 kilometer stretch.

Having brought up LIGO's exquisite sensitivity again, we need to qualify our description of what its sensitivity really means, as there is an important detail about how detectors like LIGO operate that we have not yet mentioned. The *noise* in LIGO—namely, all the other things that cause the distances along its measuring arms to vibrate and so masquerade as gravitational waves—is actually quite large, and only for a rare, very loud event, such as the black hole collision heard on September 14, 2015, will the signal be audible

CHAPTER SIX

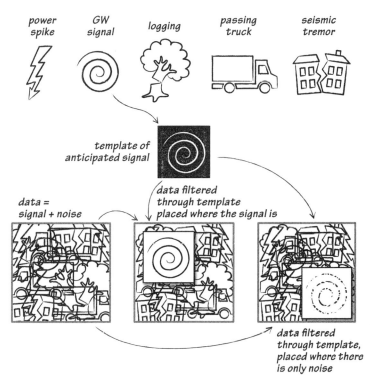

FIGURE 6.4. How templates can help uncover a signal buried in noise. When the template is centered on the signal, a clean, solid representation of the template emerges. When centered on noise, a splotchy and incomplete version results. GW stands for "gravitational wave."

above the noise. Yet LIGO can still measure subaudible signals by applying sophisticated data analysis techniques using what are called libraries of *templates:* large libraries of theoretical predictions for all the possible gravitational waves LIGO might see. Think of one template as the fingerprint for the corresponding event, for example the inspiral of two nonrotating 32-solar-mass black holes. Noise sources,

BLACK HOLE COLLISIONS

such the ground vibrations caused by a truck driving along a nearby road, or the logging activity taking place near the LIGO Livingston site, also have distinctive fingerprints, but their exact shapes are very different from the black hole merger chirp. LIGO is continually bombarded by noise, and so what LIGO measures is a jumble of all these noise-fingerprints stamped on top of one another together with, every now and then, a gravitational wave fingerprint. The template can then serve as a mask, letting through a gravitational wave fingerprint that exactly matches it, while at the same time blocking the parts of the noise-fingerprints that don't. Though this technique cannot eliminate noise completely, it helps a lot—enough to allow LIGO to listen for gravitational wave events that occur much farther away from the Earth than would otherwise be possible.

A couple of challenges related to templates had to be overcome (and are still being worked on) for LIGO to be in a position to fully realize its promise as a gravitational wave observatory when it reaches design sensitivity. The first is simply to calculate all the templates for the black hole collisions and other plausible events that could serve as sources for gravitational waves that LIGO can see. These other sources include collisions between black holes and neutron stars; mergers of two neutron stars; supernova explosions; rapidly spinning neutron stars with bumps on their surfaces;[3] a cosmic ocean of gravitational waves created early in the evolution of the universe; and various snaps, kinks, and intersection events in a network of cosmic strings that some

---

3 These bumps are analogs of mountains on Earth, but on neutron stars the "mountains" are only a few millimeters high at most because of the extremely high density of matter and strong gravitational fields at the surfaces of neutron stars.

CHAPTER SIX

cosmologists have hypothesized to exist.[4] It is hard work to solve the Einstein field equations in all these diverse settings, but doing so is necessary to build the templates needed by LIGO. Most challenging in this regard are collisions. After several decades of concerted effort by dozens of researchers, the binary black hole merger problem has largely been solved. This has been achieved through a combination of analytical methods (well suited to the early stages of an in-spiral) and numerical simulations using supercomputers (needed to model the late inspiral and collision of the two black holes). Neutron stars add new layers of complexity to the problem because the equations governing the dynamics of matter comprising neutron stars also need to be solved. In fact, as discussed in Chapter 5, we don't have detailed knowledge of the properties of nuclear matter at the exceedingly high densities thought to exist in neutron stars. Templates can be constructed that quantify our uncertainties about dense nuclear matter, and it is reasonable to hope that gravitational wave observations of merging neutron stars, or of black hole–neutron star collisions, will teach us a lot about the properties of matter at extremely high densities.

The second issue with templates is more nefarious. How do we observe gravitational wave events that we have not anticipated and so could not have constructed templates for? A related question that is just as worrisome is: What if our theoretical models of anticipated events are not quite correct? At first glance then, it seems as if LIGO is a somewhat biased scientific measurement device and may not be capable of discovering anything completely new or unexpected.

4 A cosmic string is a very thin but extremely dense stream of energy predicted by some theories, but none have yet been observed.

BLACK HOLE COLLISIONS

However, things are not as dire as they might sound: LIGO does employ template-free analysis methods that would notice a gravitational wave that is sufficiently loud, even if its form doesn't match anything in the template database. Similarly, if a passing gravitational wave only partly matches a template because something in the template isn't quite correct, the same template-free analysis would notice what's called a *residual*: a signal left over after subtracting out the best-fit template that is inconsistent with noise alone. So the bottom line is that templates allow us to hear more distant events than would otherwise be possible and to match these signals up with predicted sources, but they don't preclude the possibility that LIGO will discover anomalous gravitational wave signals whose mysterious origin would then need to be understood.

We end this chapter by discussing numerical simulations of colliding black holes. Such simulations are a crucial ingredient to the creation of the templates that LIGO uses. We should emphasize that using numerical methods to solve equations is often a last resort, which we turn to only when all pencil-and-paper calculations are judged insufficient to get at the physics that we want. Also, we restrict our discussion to collisions of black holes in the absence of matter, so that the equations we must solve are the Einstein field equations in vacuum, $G_{\mu\nu} = 0$. How can this be so difficult?

The difficulty is that Einstein's field equations are differential equations, which means that they are framed locally in terms of the way the metric varies across an infinitesimally small region of space and time. Differential equations are almost always hard to solve, and vast swaths of research across math, physics, chemistry, and engineering are devoted to methods of finding approximate solutions to differential equations. Think of it like this. A computer can be told how to add, subtract, multiply, and divide, and it can do these basic arithmetic operations extremely quickly. But a true solution of a differential equation involves, in principle, infinitely many basic operations, because the answer is not just a number, but some continuous curve—or curved spacetime in the case of Einstein's equations—that requires infinitely many numbers to specify. Of course, no computer can do infinitely many calculations in finite time. So what we want instead is some strategy for doing finitely many calculations which nevertheless comes very close to the exact solution of the differential equations of interest. More precisely, we want a strategy for producing a sequence of approximate digital solutions, each giving a closer approximation to an exact solution of the differential equations than the one before it. We do not consider a numerical problem solved until we have good evidence that our approximate digital solutions can be made as close to an exact solution as we like, given sufficient computer time. As an analogy, think of viewing a video over a slow internet connection. If your browser software is well configured, you'll get a blurry, blocky version of the video which nevertheless plays at the correct speed and has most of the right colors and some of the right shapes. If the connection is faster, or you wait a bit longer for more of the video to download before starting to play, the computer will refine the blocks of each image into smaller blocks that show more accurate detail and less blurry colors. If you wait to download the entire video, or if you have a fast connection, you'll see the video play with the maximum resolution available. Successive digital approximations to differential equations are like that, but a difference is that in principle, there is no limit to how much you can improve the approximation. The only limit is how much computing time you are willing to allocate to the task. Numerical "simulation" of Einstein's equations is almost a misnomer, because it suggests that the goal is to imitate the essential features of curved spacetime while ignoring some of the details. In truth, the goal is to have a strategy for mapping out all the details of spacetime to within any prespecified accuracy in finite time. An additional hallmark of a successful numerical simulation of colliding black holes is that you can achieve a good match to the approximate accounts of the inspiral and ringdown regimes that were available before the modern methods of numerical simulation reached maturity.

What sort of strategy should we employ to numerically simulate the vacuum Einstein field equations? Let's think about what the answer should look like. We want a digital representation of the metric describing the geometry of spacetime. Recall that a metric is a rule for deciding the distance between any two points. Differential geometry allows us to restrict attention to points which are close together. The so-called metric tensor tells us the distance from a given point to any other sufficiently close point. Practically speaking, the metric tensor is a four-by-four matrix of numbers. Having an exact solution of the field equation means that we know the metric tensor exactly at every point in spacetime. The Schwarzschild and Kerr black hole solutions provide that exact information in terms of some very clever mathematical formulas. In a numerical simulation, exact formulas aren't available, and of course we can't specify the metric tensor at the infinitely many points of spacetime. So what we do is isolate a spacetime region of primary interest (say, some region surrounding a pair of black holes which are about to merge) and fill it with a grid of points. At each of these finitely many grid points we want to assign an approximate value of the metric tensor. Just as with the analogy of a slowly downloading video, our aim is to be able to refine the grid over and over, and with each refinement we want to be able to give more and more accurate values for all components of the metric at each point on the grid. In short, we are discretizing curved spacetime so as to reduce it to a mathematical construct that a computer can deal with. Discretizing on ever finer grids is the keystone of our numerical simulation strategy. Typical large simulations today can have hundreds of millions or even billions of grid points.

Imposing the vacuum Einstein equations means that spacetime cannot curve any which way, but only according to certain constraints which tell us how the metric pushes and pulls on nearby bits of itself. The true Einstein equations are differential equations, which means that "nearby" is understood as "arbitrarily close." When we deal with discretized spacetime, we have to alter the Einstein equations a little bit so that they become rules for how the metric at a given point pushes and pulls on the metric at neighboring points on the grid.[5] At least in principle, these discretized Einstein equations can be handled by a computer because they involve only finitely many equations in finitely many variables.

There remain two difficulties which are peculiar to general relativity: singularities and constraints. The problem of singularities is actually a familiar one, and entirely physical:

5 Paradoxically, a useful strategy in numerical work is to allow the pushing and pulling to extend over several or even many lattice spacings rather than just one or two. Conceptually it is easiest to think about nearest-neighbor interactions only.

BLACK HOLE COLLISIONS

Black holes hide singularities at which the Einstein field equations become nonsensical. If we're not careful, numerical simulations of spacetime will tend to explore the interior of black holes, and when the computer runs into a singularity, there's a problem. This might seem like a minor matter, since physical intuition tells us that any problems that the computer encounters inside a black hole horizon can be ignored because no signals can come out to "pollute" the rest of the simulation. In fact, the real story is more subtle. If a singularity is encountered at some lattice point—meaning that the metric tensor contains some infinite components—then the push-pull encoded in the discretized Einstein equations will make neighboring lattice points singular as well. Then their neighbors will become singular, and so forth. It is hard to design a code which keeps singularities from spreading uncontrollably. The correct approach is to identify a horizon soon after it forms and to instruct the computer not to look too deeply inside the horizon. By keeping a small layer of spacetime inside the horizon, one can ensure that all the near-horizon physics of classical relativity is properly captured by the discretized Einstein equation; but by excising the deep interior we keep the computer from encountering a singularity. This excision strategy makes strong use of Penrose's Cosmic Censorship Conjecture, according to which singularities in solutions to Einstein's equations will never arise except inside event horizons. The fact that numerical simulations of Einstein's equations work when we implement an excision strategy as just described is actually impressive evidence in favor of Cosmic Censorship.

The problem of constraints is of a more technical nature, but it is worthy of mention because it figures prominently in the way numerical simulation of the Einstein equations

CHAPTER SIX

is really done. Usually we start with some initial geometry, like two nonspinning black holes in orbit around each other, and ask what happens as we move forward in time. What that means in practice is that we think of our big grid discretizing four-dimensional spacetime as divided up into three-dimensional spatial slices, and we define the flow of time in terms of the lapse function to connect these slices together. The common term for each three-dimensional spatial slice is "time-slice," because we think of it as the set of points at a definite time. What we want is to tell our computer the metric on just a few successive time-slices (maybe just two), and then ask it to advance to the next time-slice using the discretized Einstein equations. We plan to repeat this procedure, augmented by an excision strategy to avoid singularities, for as long as we need in order to see the black holes merge. And we expect that the geometry will evolve until, in the last time-slice of our simulation, we see the single merged black hole plus a snapshot of all the gravitational waves produced in the collision streaming outward from it.

Now, here is where the rot sets in. Once a choice of time-slices is made, it turns out that some of the Einstein equations don't help us in advancing from one time-slice to the next; instead, they amount only to constraints on what kind of geometries are allowed on each time-slice. Even if we carefully arrange to satisfy the constraints perfectly on one time-slice, we usually find that, using the discretized Einstein equations to evolve forward in time, the constraints become imperfectly satisfied on the next time-slice. Worse, these imperfections tend to grow with time, turning the output of our simulations into garbage! The solution is as subtle as the problem. Instead of trying to satisfy the constraints in each time-slice perfectly, we anticipate that they won't be exactly satisfied, but we alter the discretized Einstein equations by adding what can loosely be thought of as a restoring force that pushes the solution back toward one satisfying the constraints. This restoring force acts much like the restoring force that a spring can exert: Pull a spring out of equilibrium and it will pull back to try to return to equilibrium, with more force the farther it's been displaced from equilibrium. With the Einstein equations we're not adding any physical force—it's more a mathematical trick that accomplishes something similar, where "equilibrium" is a solution satisfying the constraints. An appropriate treatment of constraints along the lines just described, together with carefully selected choices on how to represent Einstein's equations on discretized spacetime, leads to simulations that indeed are able to capture all the details of spacetime in black hole collision events—provided of course we ask questions only about the geometry outside the horizon.

To sum up, most black hole collisions in the universe are probably of the inspiral-and-merge variety, which we can describe by numerically simulating the vacuum Einstein field equations $G_{\mu\nu} = 0$. Doing so for a large variety of initial conditions lets us see what kind of gravitational radiation boils off black holes as they merge. The release of energy is astonishingly fast in these processes—so fast that the gravitational luminosity of merging black holes can briefly exceed the ordinary luminosity of all the stars in the universe combined. Ordinary luminosity is starlight. Gravitational luminosity is gravitational radiation, which propagates outward from merging black holes and can be detected by L-shaped gravitational wave detectors like LIGO. In the future, we hope that gravitational waves will reveal as much about the universe as visible light has done. Gravitational waves from

CHAPTER SIX

neutron star mergers could be the next big discovery. Gravitational waves from the very early universe, if detected, could help tell us what the universe was like at the earliest times. Best of all would be to discover gravitational waves of a sort that no one anticipated! Then theorists would be put to the test to figure out what exotic physical processes produced them.

CHAPTER SEVEN

BLACK HOLE
THERMODYNAMICS

So far, we have thought of black holes as astro-physical objects that are created in supernovas or that reside in the centers of galaxies. We see their effects indirectly by observing the acceleration of nearby stars. The famous observation by LIGO of gravitational waves on September 14, 2015 is a more direct observation of a black hole collision. The tools we need to understand black holes in these contexts are differential geometry, Einstein's equations, and powerful analytical and numerical methods to solve Einstein's equations and describe the spacetime geometries that black holes create. From an astrophysical point of view, we could consider the subject of black holes closed once we can give a fully quantitative account of the relevant spacetimes. From a broader theoretical perspective, there is much more to explore. The purpose of this chapter is to hit some highlights of modern theoretical developments in black hole physics, in which ideas from thermodynamics and quantum theory intersect with general relativity to produce some surprising new insights. The upshot is that black holes are more than just geometrical objects. They have a temperature, they have enormous entropy, and they may be manifestations of quantum entanglement. Our account of the thermodynamic and quantum aspects of black holes will be more sketchy than our discussion in previous chapters of the purely geometrical features of black hole spacetimes. But quantum aspects in particular are an essential and vital part of ongoing theoretical research on black holes, and we are eager to indicate at least the flavor of this work.

In classical general relativity—meaning the differential geometry of solutions to Einstein's equations—black holes are truly black, in the sense that nothing can escape from them. Stephen Hawking showed that this situation changes completely when we include quantum effects: Black holes in fact emit radiation of a definite temperature known as the Hawking temperature. For astrophysically sized black holes (i.e., stellar mass to supermassive black holes), the Hawking temperature is negligibly small compared with the temperature of the cosmic microwave background, which is a form of radiation that fills the universe and can itself be regarded as a variant of Hawking radiation. Hawking's calculation of black hole temperature is part of a larger program of research called black hole thermodynamics. Another big part of that program is black hole entropy, which characterizes the amount of information that is lost inside the black hole. Ordinary objects (such as a cup of water, a bar of pure magnesium, or a star) also have entropy, and one of the punch lines of black hole thermodynamics is that a black hole of a given size has more entropy than any other form of matter

BLACK HOLE THERMODYNAMICS

that can be fit into a region of the same size without creating a black hole.

Before we delve deeply into Hawking radiation and black hole entropy, let's take a quick tour of quantum mechanics, thermodynamics, and entanglement. Quantum mechanics was developed mostly in the 1920s, and its initial aim was to describe very small things like atoms. Long-treasured notions such as the exact positions of individual particles become blurred in quantum mechanics, so that for example the electrons around an atomic nucleus cannot be located exactly. Instead, electrons move in so-called orbitals in which the actual locations of the electrons can only be determined probabilistically. But for our purposes, it is crucial not to jump to probability too quickly. A hydrogen atom (to take a simple example) can be in a definite quantum state. The simplest state of a hydrogen atom is its ground state, which is the state of lowest energy, and its energy is known exactly. More generally, quantum mechanics permits us (in principle) to be absolutely certain about the state of any quantum system.

Probability enters when we ask certain questions about a quantum mechanical system. For example, if a hydrogen atom is definitely in its ground state, we could ask, "Where will the electron be found?" and the rules of quantum mechanics would return only a probabilistic answer, which roughly comes down to "The electron is probably within half an angstrom of the hydrogen nucleus." (An angstrom is 10^{-10} meters.) Now, it's possible through a physical process to find out the position of the electron much more precisely than an angstrom. A typical process is to scatter a photon with very short wavelength off the electron, after which we can reconstruct where the electron was at the instant of scattering to within approximately one wavelength of the photon. This physical process changes the state of the electron, so that it no longer is in the ground state of the hydrogen atom and no longer has definite energy. Instead, temporarily, it has almost a definite position (meaning, to within about one wavelength of the photon we used). Which position it has can only be anticipated probabilistically to within about an angstrom, but once we measure it, we will know what we found. In short, if we measure a quantum mechanical system in some way, then at least in the conventional view we are forcing it into a state with a definite value for the quantity we are measuring.

Quantum mechanics applies not just to small systems, but (we think) to all systems. However, the rules of quantum mechanics quickly become complicated for larger systems. A key feature is the notion of quantum entanglement. A simple example of entanglement can be illustrated using the notion of spin. Individual electrons have spin, so in practice a single electron can have spin up or spin down with respect to a chosen spatial axis. The spin of an electron is observable because the electron produces a small magnetic field similar to the magnetic field from a bar magnet, and spin up means that the north pole of the electron points down while spin down means the north pole points up. Two electrons can be prepared in a joint quantum state in which one of them has spin up while the other has spin down, but it is impossible to tell which electron has which spin.[1] In fact, the ground state

---

1 Informed readers might be bothered by the fact that electrons are identical particles. None of our discussion requires the special properties of wave-functions that arise for identical particles. We could, for example, take an electron and a proton to form our combined quantum state, and still we would not be able to say which particle was spin up and which spin down.

BLACK HOLE THERMODYNAMICS

of a helium atom involves two electrons in precisely such a state, which is called a spin singlet state because the combined spin of the two electrons is zero. If we separate the two electrons without disturbing their spins, then we can continue to assert that they are jointly in a spin singlet state, but we can't say what the spin is of either electron by itself. If we measure one spin and find that it is up, then we can be perfectly confident that the other spin is down. We say in this situation that the spins are *entangled*, because neither spin by itself has a definite value, but together they are in a definite quantum state.

Einstein was deeply troubled by entanglement because it seems to defy principles of relativity. Consider the case of two electrons in a spin singlet state which are widely separated in space. To be definite, let one electron be given to Alice and the other to Bob. Let's say Alice measures her electron's spin and finds that it is spin up; but Bob refrains from any measurement whatsoever. Before Alice does her measurement, it is impossible to say what the spin of Bob's electron is. But the moment she completes her measurement, she can say with perfect confidence that Bob's electron has spin down (the opposite of what she found for her spin). Does this mean that her measurement instantaneously forced Bob's electron into the spin down state? How could that happen if the electrons are spatially separated? Einstein and his collaborators, Nathan Rosen and Boris Podolsky, felt that the issues surrounding measurement of entangled systems were so serious that they jeopardized quantum mechanics itself. The Einstein-Podolsky-Rosen (EPR) paradox uses a setup like the one we've just described to claim that quantum mechanics cannot be a complete description of reality. Modern consensus, based on further theoretical work and many measurements, is that the EPR paradox is invalid, and quantum theory is right. Quantum mechanical entanglement is real, and a symptom of it is that measurements on entangled systems will be correlated even if they are widely separated in spacetime.

Let's go back to the situation where we prepare two electrons in a spin singlet and give one each to Alice and Bob. Before any measurements take place, what can we say about the electrons? Together, they are in a definite quantum state (the spin singlet state). Alice's spin by itself is equally probable to be spin up or spin down. More precisely, its quantum state is equally likely to be the spin up quantum state or the spin down quantum state. We are now relying on probability in a more profound way than before. Previously, we considered a definite quantum state (the ground state of hydrogen) and learned that there are certain "bad" questions, like "Where will the electron be found?" which have probabilistic answers. If instead we had asked "good" questions, like "What is the energy of the electron?" we would have gotten definite answers. Now, there are no "good" questions to ask about Alice's electron without reference to Bob's. (We're excluding silly questions like "Does Alice's electron have spin?" which are constructed so as to have only one possible answer.) Thus, in discussing half an entangled state, we must use probabilities to characterize how things are. Certainties arise only when we consider how the answers to questions Alice and Bob can ask are related.

We started on purpose with one of the simplest quantum mechanical systems we know: the spins of individual electrons. Quantum computers, we hope, will be built from similarly simple systems. In fact, the spins of individual electrons, or other equivalent quantum systems, are now described as

BLACK HOLE THERMODYNAMICS

qubits (short for "quantum bits") in anticipation of their role in quantum computers analogous to that of bits in digital computers.

Now suppose we replace each electron by a far more complicated quantum system with many quantum states rather than just two. For example, perhaps we give Alice and Bob blocks of pure magnesium. Before Alice and Bob go their separate ways, their blocks are allowed to interact, and we prescribe that they start in a definite joint quantum state. Once Alice and Bob separate, their magnesium blocks no longer interact. Just as for electrons, each magnesium block is in an uncertain quantum state, even though considered jointly the blocks are in a definite quantum state. (This discussion assumes that Alice and Bob are able to move the magnesium blocks without disturbing each block's internal state in any way, just as we assumed earlier that Alice and Bob could separate the entangled electrons without disturbing the electrons' spins.) What's different now is that the uncertainty about the quantum state of each block by itself is enormous. Each block could easily have more quantum states accessible to it than there are atoms in the universe. This is where thermodynamics enters in. A very imprecisely specified system may nevertheless have some macroscopic properties that are well defined. Temperature is such a property. Temperature is a measure of how likely any part of a system is to have a certain average energy, with higher temperatures corresponding to a high likelihood of large energies. Entropy is another thermodynamic property, and it is essentially the logarithm of the number of states available to a system. Yet another thermodynamic property that would be interesting for a block of magnesium is the total magnetization, which essentially

CHAPTER SEVEN

amounts to how many more electrons inside the block are spin up than are spin down.

We have introduced thermodynamics as a way of treating a system whose quantum state is not precisely known due to some entanglement with another system. This is a really powerful viewpoint, but it is far from the way that the originators of thermodynamics thought. These people, such as Sadi Carnot, James Joule, and Rudolf Clausius, worked in the midst of the industrial revolution of the nineteenth century, and they were interested in the most practical of questions: How do engines work? Pressure, volume, temperature, and heat were the bread and butter of engine design. Carnot established that energy supplied in the form of heat could never be converted entirely into useful work, like lifting a heavy object. There would always be some waste. Clausius made a key contribution by introducing the idea of entropy as a uniform accounting tool that measured how much waste was generated by any heat-related process. The key insight was that entropy can never decrease, and in almost all processes it increases. The processes where entropy increases are called irreversible, precisely because they cannot happen in reverse without decreasing entropy. The subsequent development of statistical mechanics by Clausius, Maxwell, and Ludwig Boltzmann (among others) showed that entropy is a measure of disorder. Usually, the more you push at something, the more disordered it gets. If you design a process to create order, it will inevitably generate more entropy than it destroys, for instance by the release of heat. For example, a crane which stacks steel bars in perfect alignment is creating order as far as the positioning of the steel bars is concerned, but all the requisite lifting will generate heat as a by-product, so that overall entropy increases.

BLACK HOLE THERMODYNAMICS

The nineteenth-century viewpoint on thermodynamics seems further from the quantum entanglement perspective than it really is. Any time a system interacts with an external agent, its quantum state becomes entangled with the quantum state of the external agent. Usually, that entanglement will lead to greater uncertainty about the quantum state of the system, or in other words an increase in the number of quantum states available to the system. As a result, entropy of the system—as defined in terms of the number of available quantum states—usually increases as a result of interactions with other systems.

To summarize, quantum mechanics provides a new way of characterizing states of physical systems in which some quantities (like position) become blurry, but others (for instance, energy) often are known precisely. In quantum entanglement, two in-principle separate systems have an overall quantum state which is known, but each part by itself has an uncertain quantum state. The standard example of entanglement is a pair of spins in a singlet state, where it is impossible to know which spin is up and which is down. Uncertainty in the quantum state of a large system leads to the study of thermodynamics, where macroscopic properties like temperature and entropy are known to good accuracy despite there being many possible microscopic quantum states of the system.

Having completed our quick tour of quantum mechanics, entanglement, and thermodynamics, let's start along the road to understanding how they are used to show that black holes have a temperature. A first step was taken by Bill Unruh, who showed that an accelerating observer in flat space would perceive a temperature equal to his acceleration divided by 2π. The key to Unruh's calculation is that an

CHAPTER SEVEN

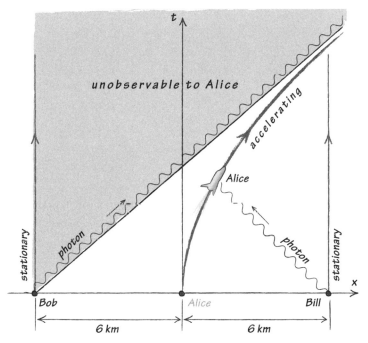

FIGURE 7.1. Alice accelerates from rest as Bob and Bill remain stationary. Alice's acceleration is just enough so that she never receives the photons that Bob sends out toward her when he waves at time $t = 0$. She does however receive photons sent toward her by Bill at $t = 0$. The upshot is that Alice is able to observe only half of spacetime.

observer moving with constant acceleration in a particular direction can only see half of flat spacetime. In effect, the other half is behind a horizon that is similar to a black hole horizon. This sounds impossible at first. How can flat space-time exhibit a black hole horizon? To get the idea, let's bring in our reliable troupe of observers, Alice, Bob, and Bill. At our request, they line up with Alice between Bob and Bill, and with 6 kilometers of separation between adjacent pairs

BLACK HOLE THERMODYNAMICS

of observers. It is agreed that at time 0, Alice will jump into a rocket ship and start moving toward Bill (and so away from Bob) with constant acceleration. Her rocket ship is a very good one, capable of an acceleration of 1.5 trillion times the gravitational acceleration experienced by an object near the surface of the Earth. This is obviously tough on Alice; but, as we'll see, the numbers are chosen with a definite purpose in mind—and after all we're discussing conceptual possibilities only. Right at the moment when Alice jumps into her rocket ship, Bob and Bill both wave to her. (We can give meaning to the words "right at the moment" because before Alice starts moving, she is in the same frame of reference as Bob and Bill, so they can all agree on a definite notion of time.) Alice will certainly see Bill wave: In fact, she will see him wave sooner in her rocket ship than she would have if she had stayed where she was, because in her rocket ship she is moving toward his position. In contrast, she is moving away from Bob's position, so we might reasonably conclude that she sees him wave somewhat later than she would have if she had stayed where she was. The truth is more remarkable: She will never see Bob wave! To put it another way, the photons that Bob transmits toward Alice while waving will never catch up with her, even though she never quite reaches the speed of light. If Bob had started somewhat closer to Alice, then the photons he sends at the moment of her departure would have caught up with her. If he had started any farther away, then they wouldn't have. It is in this sense that Alice can see only half of spacetime. As of the moment Alice starts moving, Bob is standing a little bit behind the horizon that Alice observes.

In our discussion of quantum entanglement, we got used to the idea that even if a quantum mechanical system as a

CHAPTER SEVEN

whole has a definite quantum state, parts of it may not. In fact, when we are discussing a complicated quantum system, a part of it may be best characterized in thermodynamic terms, with a definite temperature despite its highly uncertain quantum state. Our current setup with Alice, Bob, and Bill is a bit like that, but the quantum system we're thinking of is empty spacetime. Alice can see only half of it. Let's stipulate that spacetime as a whole is in its ground state, meaning that there are no particles present (aside from Alice, Bob, Bill, and the rocket ship). The part of spacetime which Alice can see will not be in its ground state; instead, it will be in an entangled state with the part she can't see. By itself, spacetime as Alice experiences it is in a complicated, uncertain quantum state characterized by a finite temperature. Unruh's calculation shows that this temperature is about 60 nanokelvins. In short, Alice sees a thermal bath of radiation as she accelerates, and the temperature of this bath is (in appropriate units) her acceleration divided by 2π.

The uncomfortable part of Unruh's calculation is that although it refers from first to last to empty space, it defies King Lear's dictum, "Nothing will come of nothing." How can empty space be so complicated? The fact is, in quantum theory, empty space is a very busy place. Ephemeral excitations called virtual particles, with both positive and negative energy, are constantly popping into and out of existence. An observer in the far future, call her Carol, who can see essentially all of empty space, can confirm that no particles have a durable existence. The presence of positive energy particles in the part of spacetime that Alice can observe is tied through quantum entanglement with excitations of equal and opposite energy in the region of spacetime that she cannot observe. Carol perceives the whole quantum truth about empty spacetime, which is that there are no particles. Alice's experience however is that there are.

The Unruh temperature feels like a fake because it is not a property so much of flat space as a property of an observer undergoing constant acceleration in flat space. However, gravity itself is a "fake force" in the sense that the "acceleration" that it causes is nothing more than geodesic motion in a curved metric. As we explained in Chapter 2, Einstein's principle of equivalence states that acceleration and gravitation are essentially equivalent. From this point of view, it is not too shocking that a black hole horizon has a temperature equal to the Unruh temperature of an accelerating observer. But, we might ask, what acceleration should we use? If we stand sufficiently far away from a black hole, its gravitational pull on us becomes as weak as we please. Should we therefore use a correspondingly small acceleration to determine the black hole's effective temperature as measured by us? That would be an uncomfortable situation, because the temperature of an object is not supposed to decrease arbitrarily; it's supposed to be some fixed value that even a very distant observer will be able to measure as finite.

A viewpoint more in the spirit of Hawking's account of black hole temperature is that we should use the acceleration of an observer hovering very close to the black hole horizon but then discount the corresponding temperature by the gravitational redshift factor experienced by that observer. This viewpoint is the closest we'll get to the actual calculation of the Hawking temperature, so let's walk through it slowly in the case of a Schwarzschild black hole. By a hovering, or *static* observer, we mean one who stays at a fixed radius from the horizon without orbiting around the black hole. To do this, the observer, let's call her Anne, must constantly push herself away from the black hole, for instance by using a rocket. If Anne looks only at her local geometry, the equivalence principle says that she won't be able to distinguish it from flat space through which she is accelerating at a constant rate. The closer Anne is to the actual black hole horizon, the larger this apparent acceleration will be. According to Unruh's calculations, Anne will perceive a temperature equal to her acceleration divided by 2π. We seem to be falling into the same trap as before: Perceived temperature depends on position. What saves the day is that Anne is also experiencing considerable gravitational redshift compared to another observer, call him Bart, who remains far away from the black hole. (In this context, "far" means many times the Schwarzschild radius.) The closer to the horizon Anne gets, the larger will be the Unruh temperature that she perceives. But her increased gravitational redshift means that by the time the radiation she sees claws its way out of the gravitational field of the black hole and reaches Bart, it will be at a finite temperature which does not change as Anne gets closer and closer to the horizon. This finite temperature is the Hawking temperature, and multiplying it by 2π gives a quantity called the *surface gravity* of the black hole. Surface gravity is the acceleration Alice would have to undergo in flat space to see the same temperature of Unruh radiation as the Hawking radiation Bart sees.[2]

---

2 Surface gravity can be defined and calculated without any reference to temperature or quantum effects, in terms of the spatial variation of the gravitational redshift as one approaches the horizon. It turns out to equal the acceleration that would be experienced by a hypothetical observer in Newtonian gravity sitting on the surface of a sphere with the same radius and mass as that of the black hole; hence the name "surface gravity."

BLACK HOLE THERMODYNAMICS

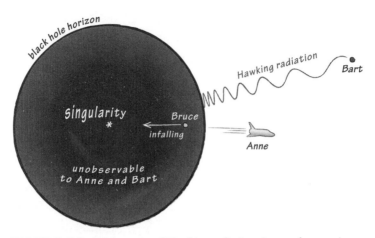

FIGURE 7.2. A schematic of Hawking radiation. Anne, who remains at a fixed radius near the horizon, is equivalent to an accelerating observer because she feels the black hole's gravitational pull. She sees radiation for reasons similar to the Unruh effect. That radiation is gravitationally redshifted as it propagates outward to Bart, who is also stationary, but he is so far from the black hole that he scarcely feels its gravity. The infalling observer, Bruce, sees no Hawking radiation as he crosses the horizon.

We said earlier that we chose numbers for our example of the Unruh effect for a definite purpose. In fact, Alice's acceleration, 1.5 trillion times the Earth's gravitational acceleration, is just about equal to the surface gravity at the horizon of a black hole whose mass equals the mass of the Sun. Correspondingly, the Hawking temperature of such a black hole is the same as the Unruh temperature that Alice experiences, namely 60 nanokelvins. Larger black holes have even smaller temperatures, inversely proportional to their mass.

Our account of Unruh temperature emphasized that an inertial observer in the far future (we called her Carol) would observe the whole quantum truth, namely that the quantum state of all spacetime is the vacuum, with no excitations whatsoever. Alice's perceived thermal state comprises positive energy excitations which are quantum mechanically entangled with negative energy excitations in the region of spacetime that she cannot perceive. It turns out that there are analogs of these statements in the case of Hawking radiation, but with some important differences. The static observer whom we named Anne is the closest analog to the accelerating observer Alice in our discussion of the Unruh effect. Recall that in most of that discussion, Bob is behind the horizon that hides half of flat spacetime from Alice. An analog of Bob in the black hole context would be a freely falling observer, call him Bruce, who falls inside the black hole horizon. Bruce is an ill-fated individual because his future certainly includes collision with the black hole singularity. However, for a large black hole, this unhappy fate may be long delayed, and it is reasonable to ask what observations Bruce makes in the meantime. The answer is that he will not observe any temperature at all as he crosses the horizon, assuming no radiation is coming into the black hole from some other source. At least locally near the black hole horizon, Bruce will say that there are no quantum excitations at all.

Anne and Bart's descriptions are different from Bruce's in that they see positive energy particles. Just as with the Unruh effect, these positive energy excitations must be quantum mechanically tied to negative energy excitations inside the black hole horizon. Recall the tricky nature of these types of arguments: Bruce doesn't actually see the negative energy excitations when he crosses inside the horizon; instead, he sees no excitations at all. The negative energy excitations inside the horizon become necessary only to give

BLACK HOLE THERMODYNAMICS

a consistent account of quantum theory that includes Anne and Bart's perspectives as outside observers. And yet, these peculiar excitations have an important physical role. They serve to decrease the total mass of the black hole to offset the energy that Anne and Bart say is radiating out from it.

The positive energy excitations coming out and the negative energy excitations going in are quantum mechanically entangled, and at least near the horizon one can argue that this entanglement serves the purpose of keeping quantum theory consistent between an infalling observer like Bruce and static observers like Anne and Bart. It is this quantum consistency through entanglement that drives the Hawking effect. Thus, subtlety upon subtlety results in something very tangible: radiation of positive energy excitations from the black hole!

A striking contrast with the Unruh effect is that it is hard to see how to arrange for an observer in the far future to see all of spacetime, since the inside of the black hole has no far future, while from the outside of the black hole we cannot see the inside. Maybe if the black hole evaporates completely, an observer who observes the whole process can be said to be in possession of the whole quantum truth about spacetime. Or instead, perhaps no observer can see the whole truth about spacetimes involving black holes, meaning that information about the quantum state is truly lost. The conundrum of how quantum evolution of the past into the future can coexist with black holes is known as the *information loss paradox*, and it is still debated today.

Altogether, the picture of Hawking radiation is that quantum mechanical excitations exit the black hole, redshifting as they do so, and they are observed by a distant observer as a bath of radiation with temperature equal to the surface gravity of the black hole divided by 2π. Meanwhile the black hole mass slowly decreases or evaporates to reflect the energy lost to the radiation. Hard questions about the quantum experience of observers following different paths in black hole spacetimes have perplexed generations of theorists, but provided we remain well outside a black hole, and provided the black hole in question is so large that it doesn't have time to fully evaporate, thermal radiation at the Hawking temperature is what we will see.

Hawking radiation is the most celebrated thermodynamic property of black holes. However, of equal importance is the Bekenstein-Hawking black hole entropy, named after Jacob Bekenstein and Stephen Hawking. Recall that entropy measures the number of quantum states available to a system. (More precisely, entropy is the logarithm of the number of available quantum states.) Its most important property is that in physical processes, it never decreases, and usually it increases. Another important property is that the entropy of two systems together cannot be larger than the sum of the entropies of the systems separately. In ordinary matter, one usually finds that entropy of a whole is the sum of the entropy of the parts. For instance, the entropy of two ordinary cups of water at room temperature is twice the entropy of one cup of water. If two systems are entangled, then their joint quantum state may be known precisely, in which case they have no entropy at all as a whole; and yet, each system by itself may have considerable entropy!

In the case of black holes, the entropy turns out to be the horizon area divided by a constant that is related to the strength of gravity. The formula is $S = \frac{A}{4G_N}$, where G_N is Newton's constant, which also appears in Einstein's equations. This formula is so central to discussions of black holes that it usually is just called the area law. Theorems of classical general relativity show that the total area of black hole horizons must increase in processes such as black hole collisions. This result is understood as being the black hole version of the second law of thermodynamics. It is important to emphasize that these theorems hold classically, which means in the absence of quantum effects such as Hawking radiation. Indeed, Hawking radiation causes black holes to slowly lose mass, which means their horizon area decreases; however, this process is extremely slow.

The area law shows that black holes are very different from ordinary thermodynamic matter. Indeed, ordinary matter usually has entropy that scales with the volume. For instance, recall that two cups of water usually have twice the entropy of one cup. We could equally well say that the entropy of water scales with the mass, since two cups of water have double the mass of one cup. Scaling of black hole entropy with area seems to show that large black holes have far less entropy than we might naively have expected based on their volume, but far more than expected based on their mass. To see how these expectations work, consider combining two black holes, each with 1 solar mass, into one larger black hole. Our discussion will be crude because we're going to ignore the blast of gravitational waves that would come out of the merger, as discussed in Chapter 6. So ignoring that, the final black hole has twice the mass of the two initial black holes. The actual entropy of the final black hole is four times the entropy of each of the initial black holes. That's more than we would expect based on the mass, because if entropy scaled with mass, the entropy of the final black hole would be only double that of one of the initial black holes. It's less than we would expect based on the volume, because naively the final black hole encloses eight times the volume of one of the initial black holes, but it has only four times as much entropy. The correct scaling arises from thinking of entropy as relating to the horizon itself, as if a new qubit's worth of entropy is added every time the horizon increases by an increment proportional to G_N.

A striking claim, developed originally by Bekenstein, is that black holes carry more entropy than any other form of matter that could occupy the same region of spacetime. A simple version of Bekenstein's argument is that for ordinary matter packed into a finite region of spacetime to have a great deal of entropy, there has to be a great deal of matter—so much that it will be threatened by gravitational collapse. Before ordinary matter can exceed the entropy of a black hole, it in fact collapses into a black hole. In this sense, black hole collapse is the most disorderly and irreversible phenomenon possible.

String theory offers a microscopic justification for the area law in certain restricted contexts, but in general we do not have a derivation of the law from first principles. However, it has been argued by Ted Jacobson that if one assumes black hole thermodynamics, in particular the area law, together with some basic notions of differential geometry, one can derive the Einstein equations which are at the heart of general relativity. Furthermore, it's known that if Einstein's equations are modified while maintaining their underlying symmetries, then the area law changes, but the Hawking calculation essentially stays the same. So black hole entropy is believed to be quite a discriminating tool in describing the dynamics of spacetime. And yet, what exactly is black hole entropy?

BLACK HOLE THERMODYNAMICS

A recent proposal was made by Juan Maldacena and Leonard Susskind to dovetail entanglement entropy more closely with black hole entropy. It goes like this. Recall the EPR paradox, where two spins were entangled and then separated, and the confusing point was that neither spin has a definite quantum state by itself even though jointly they do. Each spin is a qubit, and each one by itself has a qubit's worth of entropy. Could we imagine, at some microscopic level, that each one is a black hole, and that their entanglement manifests geometrically as a wormhole between them? There are two obvious objections to this idea. First, a black hole with only one qubit's worth of entropy is so tiny that geometry might not mean anything. Second, as discussed in Chapter 3, wormholes are not traversable. To work around these objections, let's first imagine larger systems with more available quantum states and therefore larger entropies. But let's insist that two of these larger systems, one each in the possession of Alice and Bob, are perfectly entangled, so that their joint quantum state is precisely specified. Previously we offered bars of pure magnesium as examples of larger systems, but now we want to use some more arcane state of matter which after a while collapses into a black hole. In short, Alice and Bob wind up far apart, each in the vicinity of a black hole, and at least a large part of the entropy of each black hole is due to the quantum mechanical entanglement between the two systems. Then the proposal is that a wormhole connects the black holes, and this wormhole is a geometric manifestation of their entanglement.

How would we test this idea? Well, consider a thought experiment where Alice and Bob each measure their respective systems. Looking closely at a system undergoing gravitational collapse is a risky undertaking because in all

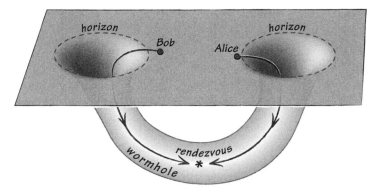

FIGURE 7.3. A wormhole connecting regions of spacetime close to Alice and Bob, respectively. Wormholes arising from entangled quantum states enable Alice and Bob to meet by both of them jumping into black holes. They can then verify their quantum entanglement before their fatal impact with the black hole singularity.

likelihood, the observer will be sucked into the black hole in the process. This sounds bad even from a conceptual point of view because it seems to exclude the possibility that Alice and Bob could make their measurements and then later compare what they found to check that their systems really were entangled. But wait! We've assumed that there's a wormhole connecting the two black holes. Alice and Bob may both be caught inside their respective black holes, but because of the wormhole, these black holes share the same interior. It's impossible for a single observer to traverse the wormhole from one exterior to the other, but it actually is possible for two observers who enter the wormhole from opposite ends to meet on the inside. So Alice and Bob could in fact compare notes. That is an important point in favor of the formation of a wormhole in the first place, because if such a wormhole did not form, then Alice and Bob really

BLACK HOLE THERMODYNAMICS

couldn't compare the results of their measurements, and the notion of quantum entanglement would be in jeopardy. All this may be scant comfort to Alice and Bob, because within a finite time of falling into their respective black holes, they will collide with the common black hole singularity. At least they could make one last test of quantum mechanics before the end!

Taking a step back from this intensely theoretical discussion, we can reasonably ask whether there is any practical interest in thought experiments of the type just described, where as part of the experiment the observers must jump into black holes. Observers who remain outside can never find out (at least by classical means) whether Alice and Bob met or not. Is it all moot? Generally the sense among theorists is that it isn't. We must keep in mind that black hole horizons are a matter of future destiny rather than momentary experience. We ourselves could at this moment be crossing into a cosmic black hole whose future singularity is farther in the future than the present age of the universe. The eventual collapse could be a preferable fate to the endless expansion of the universe envisioned by many cosmologists. Might some sort of creative destruction at the end of spacetime itself open our perceptions to vistas not yet dreamed?

CHAPTER SEVEN

EPILOGUE

We don't believe in time travel, and we're not into mysticism. But what if we could write a letter to Albert Einstein to tell him about gravity and black holes in a few paragraphs? We'd write something like this letter.

July 2017

Dear Albert,

First, you're the greatest. The one equation in physics that everyone knows is $E = mc^2$. *Time* magazine named you the Person of the Century. Einstein jokes aren't told much anymore because everyone sees the punch lines coming from a mile away. We've got a gazillion nuclear weapons, but we haven't blown ourselves up yet. In fact, the atom bombs dropped at the end of World War II are the only ones ever used to intentionally kill people.

We're really interested in general relativity and black holes these days because a big instrument called LIGO detected gravitational waves from a black hole collision that occurred over a billion years ago. We wrote a book about black holes, and since we know you were deeply interested in Schwarzschild's solution, and maybe sometimes unsure of its physical significance, we thought we would tell you about what's been happening in the sixty-plus years since you've passed on.

First, there's this thing called the event horizon in Schwarzschild's solution. If you go behind it, then you can't come back out again without exceeding the speed of light. If you remember the form of Schwarzschild's solution, there were some strange features when radius was equal to mass up to a factor of Newton's coupling. In particular, the time-time part of the metric vanishes—what we now call the lapse function. That's where the event horizon is. Schwarzschild's solution also has strange features when the radius goes to zero, and our best understanding is that these strange features mark a spacetime singularity where geometry itself breaks down. If you enter into a Schwarzschild black hole, you're bound to encounter the singularity, but we have no idea what would happen next—or even if "next" is a good word to use.

We wish you could have seen all the work done on general relativity in the twenty or so years after your passing. John Wheeler was at the center of it all. (And we knew him! He lived until 2008 and spent time with us both at Princeton.) He popularized the term "black hole" to describe Schwarzschild's solution and related metrics. A New Zealander by the name of Roy Kerr found a generalization of the Schwarzschild metric which describes a rotating black hole. It's quite a complicated metric! And it's important because it describes the end state of collapsing stars, which always have some nonzero angular momentum.

EPILOGUE

We're pretty sure now that there are lots of black holes in the universe. Like Chandrasekhar, Tolman, Oppenheimer, and Volkoff were saying in the thirties, if you put too much mass together, nothing can hold it up. It's hard to figure out the exact numbers, but if about 3 solar masses remain after a star uses up all its nuclear fuel, it's going to collapse into a black hole. What's far more amazing is that there are much bigger black holes at the middle of galaxies. The Milky Way harbors a black hole at its center which contains about 4 million solar masses. We're not pulling your leg! Modern consensus is that a lot of galaxies have much bigger black holes at their centers, maybe containing billions of solar masses. We're not sure how these black holes formed, but in the case of the Milky Way we can be sure of its existence by tracing the orbits of individual stars and seeing the effects of the black hole's gravity.

The LIGO detection of gravitational waves was just terrific. LIGO is a big Michelson interferometer, measuring 4 kilometers on a side. LIGO stands for Laser Interferometer Gravitational-Wave Observatory. Lasers are these amazing monochromatic light sources, so focused and powerful that we can use them to weld metal, but so cheap that we build them into modern record players in place of needles. We haven't got flying cars yet, but lasers are pretty nifty. Anyway— LIGO was just setting up for a serious science run, when along came this perfect gravitational wave form which they picked up almost by accident and were able to match to a template that describes the merger of two black holes, each on the order of 30 solar masses.

EPILOGUE

172

Everyone is impressed all over again with general relativity, because it's successful in describing the strong field region near the black holes where spacetime is practically getting torn to shreds, and also the far-field region where gravitational waves are the faintest whispers gliding through spacetime.

Another of your ideas that has come a long way is the cosmological constant. Even though you called it your biggest blunder, we now think that it's present as a small correction to your field equations. It's actually important at large length scales: Astronomers can't explain the recent evolution of the expanding universe unless a whopping 70% of all energy in the universe comes from the cosmological constant, or at least something we call "dark energy" that behaves very much like it. Rather than keeping the universe static as you had hoped when you first introduced the cosmological constant (brace yourself), dark energy is starting to accelerate the universe onto a track of exponential expansion. Going in another direction, the quest for a unified theory has led to intense study of spacetimes with a negative cosmological constant. General relativity in five dimensions with a negative cosmological constant connects naturally with a quantum theory on the four-dimensional boundary of spacetime. It's almost as if quantum theory is a projection of general relativity!

We're really sure now that quantum theory is right. (Sorry about that.) A British physicist named Stephen Hawking showed that quantum theory implies that black holes emit radiation, albeit at very low temperatures. Black holes also have a tremendously large entropy, despite being almost unique as solutions to your field equations. If it makes you feel any better, the paper you wrote with Podolsky and Rosen turned out to be important. People are even trying to build quantum mechanical computers nowadays using ideas related to that paper.

A lot of Princeton professors don't wear ties to work anymore, but most of us do wear socks. Lake Carnegie is as beautiful as ever. We don't see many sailors out there, but there's been an eagle nesting right on the edge of the lake. We haven't figured out a unified field theory yet, but we're still trying. The best is yet to come.

Yours truly,
Steve and Frans

EPILOGUE

INDEX

INDEX

INDEX